Michael T. Battista

COGNITION-BASED ASSESSMENT & TEACHING
of Geometric Shapes

Building on Students' Reasoning

HEINEMANN
Portsmouth, NH

Heinemann
361 Hanover Street
Portsmouth, NH 03801–3912
www.heinemann.com

Offices and agents throughout the world

© 2012 by Michael T. Battista

All rights reserved. No part of this book may be reproduced in any form or by any electronic or mechanical means, including information storage and retrieval systems, without permission in writing from the publisher, except by a reviewer, who may quote brief passages in a review, and with the exception of reproducible pages (identified by the *Cognition-Based Assessment and Teaching of Geometric Shapes* copyright line), which may be photocopied for classroom use only.

"Dedicated to Teachers" is a trademark of Greenwood Publishing Group, Inc.

Library of Congress Cataloging-in-Publication Data
Battista, Michael T.
 Cognition-based assessment and teaching of geometric shapes : building on students' reasoning / Michael T. Battista.
  p. cm.
 Includes bibliographical references.
 ISBN-13: 978-0-325-04351-7
 ISBN-10: 0-325-04351-5
 1. Shapes—Study and teaching (Elementary). 2. Geometry—Study and teaching (Elementary).
3. Cognitive learning. I. Title.
QA462.B25 2012
372.7′2—dc23   2012017156

Editor: Katherine Bryant
Production: Victoria Merecki
Interior and cover designs: Monica Ann Crigler
Website developer: Nicole Russell
Typesetter: Publishers' Design & Production Services, Inc.
Manufacturing: Steve Bernier

Printed in the United States of America on acid-free paper

21 20 19 18 17   VP   3 4 5 6 7

Contents

Acknowledgments v

Introduction vii

Chapter 1 1
Introduction to Reasoning About Geometric Shapes

Chapter 2 8
Levels of Sophistication in Student Reasoning: Geometric Shapes

Chapter 3 65
Instructional Strategies for Geometric Shapes

Appendix 111
CBA Assessment Tasks for Geometric Shapes

Glossary 143

References 147

Acknowledgments

I would like to thank the numerous students, parents, teachers, school districts, and research assistants who participated in the CBA project.

I especially want to thank Kathy Battista who has provided invaluable advice and work throughout the CBA project.

Research and development of CBA was supported in part by the National Science Foundation under Grant Numbers 0099047, 0352898, 0554470, and 0838137. The opinions, findings, conclusions, and recommendations, however, are those of the author and do not necessarily reflect the views of the National Science Foundation.

—Michael Battista

Introduction

Traditional mathematics instruction requires all students to learn a fixed curriculum at the same pace and in the same way. At any point in traditional curricula, instruction *assumes* that students have already mastered earlier content and, based on that assumption, specifies what and how students should learn next. The sequence of lessons is fixed; there is little flexibility to meet individual students' learning needs. Although this approach appears to work for the top 20 percent of students, it does not work for the other 80 percent (Battista, 1999, 2001a). And even for the top 20 percent of students, the traditional approach is not maximally effective (Battista, 1999, 2001a). For many students, traditional instruction is so distant from their needs that each day they make little or no learning progress and fall farther and farther behind curriculum demands. In contrast, Cognition-Based Assessment (CBA) offers a cognition-based, needs-sensitive framework to support teaching that enables *all* students to understand, make personal sense of, and become proficient with mathematics.

The CBA approach to teaching mathematics focuses on deep understanding and reasoning, within the context of continually assessing and understanding students' mathematical thinking, then building on that thinking instructionally. Rather than teaching predetermined, fixed content at times when it is inaccessible to many students, the CBA approach focuses on maximizing *individual student progress, no matter where students are in their personal development*. As a result, you can move your students toward reasonable, grade-level learning benchmarks in maximally effective ways. Designed to work with any curriculum, CBA will enable you to better understand and respond to your students' learning needs and help you choose instructional activities that are best for your students.

There are six books in the CBA project:

- *Cognition-Based Assessment and Teaching of Place Value*
- *Cognition-Based Assessment and Teaching of Addition and Subtraction*
- *Cognition-Based Assessment and Teaching of Multiplication and Division*
- *Cognition-Based Assessment and Teaching of Fractions*
- *Cognition-Based Assessment and Teaching of Geometric Shapes*
- *Cognition-Based Assessment and Teaching of Geometric Measurement*

Any of these books can be used independently, though you may find it helpful to refer to several because the topics covered are interrelated.

Critical Components of CBA

The CBA approach emphasizes three key components that support students' mathematical sense making and proficiency:

- clear, coherent, and organized research-based descriptions of students' development of meaning for core ideas and reasoning processes in elementary school mathematics;
- assessment tasks that determine how each student is reasoning about these ideas; and
- detailed descriptions of the kinds of instructional activities that will help students at each level of reasoning about these ideas.

More specifically, CBA includes the following essential components.

Levels of Sophistication in Student Reasoning

For many mathematical topics, researchers have found that students' development of mathematical conceptualizations and reasoning can be characterized in terms of "levels of sophistication" (Battista, 2004; Battista and Clements, 1996; Battista et al., 1998; Cobb and Wheatley, 1988; Fuson et al., 1997; Steffe, 1988, 1992; van Hiele, 1986). Chapter 2 provides a framework that shows the development of students' thinking and learning about geometric shapes in terms of such levels. This framework describes the "cognitive terrain" in which students' learning trajectories occur, including:

- the levels of sophistication that students pass through in moving from their intuitive ideas and reasoning to a more formal understanding of mathematical concepts;
- cognitive obstacles that students face in learning; and
- fundamental mental processes that underlie concept development and reasoning.

Figure 1 sketches the cognitive terrain that students must ascend to attain understanding of geometric shapes. This terrain starts with students' preinstructional reasoning about geometric shapes, ends with a formal and deep understanding of geometric shapes, and indicates the cognitive plateaus reached by students along the way. Not pictured in the sketch are sublevels of understanding that may exist at each plateau level. Note that students may travel slightly different trajectories in ascending through this cognitive terrain, and they may end their trajectories at different places, depending on the curricula and teaching they experience.

Figure 1 Levels of Sophistication Plateaus and Two Learning Trajectories for Geometric Shapes

STUDENTS UNDERSTAND AND CREATE FORMAL DEDUCTIVE PROOFS

Learning Trajectory for Student 1 – – – –
Learning Trajectory for Student 2 ———

STUDENTS ATTEMPT TO IDENTIFY SHAPES AS VISUAL WHOLES

A Note About the Student Work Samples

Chapter 2 includes many examples of students' work, which are invaluable for understanding and using the levels. All of these examples are important because they show the rich diversity of student thinking at each level. However, the first time you work through the materials, you may want to read only a few examples for each type of reasoning—just enough examples to comprehend the basic idea of the level. Later, as you use the assessment tasks and instructional activities with your students, you can sharpen your understanding by examining additional examples both in the level descriptions and in the level examples for each assessment task.

Assessment Tasks

The Appendix contains a set of CBA assessment tasks that will enable you to determine your students' mathematical thinking and precisely locate students' positions in the cognitive terrain for learning that idea. These tasks not only assess exactly what students can do, they also reveal students' reasoning and underlying mathematical cognitions. The tasks are followed by a description of what each level of reasoning might look like for each assessment task. These descriptions will help you pinpoint your students' positions in the cognitive terrain of learning.

Using CBA assessment tasks to determine which levels of reasoning students are using will help you pinpoint students' learning progress, know where students should proceed next in constructing meaning and competence for the idea, and decide which instructional activities will best promote students' movement to higher levels of reasoning. It can also help guide your questions and responses in classroom discussions and in students' small-group work. The CBA website at www.heinemann.com/products/E04351.aspx includes additional assessment tasks that you can use to further investigate your students' understanding of geometric shapes.

Instructional Suggestions

Chapter 3 provides suggestions for instructional activities that can help students progress to higher levels of reasoning. These activities are designed to meet the needs of students at each CBA level. The instructional suggestions are not meant to be comprehensive treatments of topics. Instead, they are intended to help you understand what kinds of tasks may help students make progress from one level or sublevel to the next higher level or sublevel.

Using the CBA Materials

Determining Students' Levels of Sophistication

There are several ways to use CBA assessment tasks to determine students' levels of sophistication in reasoning about geometric shapes.

Individual Interviews

The most accurate way to determine students' levels of sophistication is to administer the CBA assessment tasks in individual interviews with students.[1] For many students, interviews make describing their thinking much easier—they are perfectly capable of describing their thinking orally but have difficulty doing it in writing. Individual interviews also enable teachers to ask probing questions at just the right time, which can be extremely helpful in revealing students' thinking. (Beyond assessment purposes, the individual attention that students receive in individual assessment interviews provides students with added motivation, engagement, and learning.)

Whole-Class Discussion

In an "embedded assessment" model—in which assessment is embedded within instruction—you can give an assessment task to your whole class as an instructional activity. Each student should have a student sheet with the task on it. Students do all their work on their student sheets and describe in writing how they solve the task. When all the students have finished writing their descriptions of their solution methods, lead a class discussion of those methods. For instance, many teachers have a number of individual students present their solutions on an overhead projector or a document-projection device. As students describe their thinking, ask questions that encourage students to provide the detail you need to determine what levels of reasoning they are using. Also, at times, you can summarize students' thinking in ways that model good explanations (but be sure to provide accurate descriptions of what students say rather than formal versions of their reasoning). After each different student explanation, you can ask how many students used the strategy described. It is

---

[1] For helpful advice on scheduling and conducting student interviews, see Buschman (2001).

important that you not only have students orally describe their solution strategies but that you talk about how they can write and represent their strategies on paper. For instance, after a student has orally described his strategy, ask the class, "How could you describe this strategy on paper so that I would understand it without being able to talk to you?"

Another way to see if students' written descriptions accurately describe their solution strategies is to ask students to come up to your desk and tell you individually what they did, which you can then compare to what they wrote.

Individual and Small-Group Work

You can also determine the nature of students' reasoning by circulating around the room as students are working individually or in small groups on CBA assessment tasks or instructional activities. Observe student strategies and ask students to describe what they are doing as they are doing it. Seeing students actually work on problems often provides more accurate insights into what they are doing and thinking than merely hearing their explanations of their completed solutions (which sometimes do not match what they did). Also, as you talk to and observe students during individual or small-group problem solving, for students who are having difficulty accurately describing their work, write notes to yourself on students' papers that tell you what they said and did (these notes are descriptive, not evaluative).

The Importance of Questioning

Keep in mind that the more students describe their thinking, the better they will become at describing that thinking, especially if you guide them toward providing increasingly accurate and detailed descriptions of their reasoning. For instance, consider a student attempting to identify the given shape.

Task *What kind of shape is this? How do you know?*

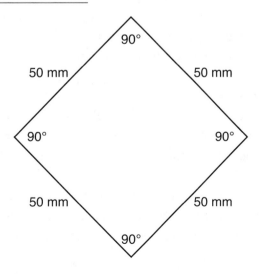

Introduction

Student: *It's a square.*

Teacher: *How do you know it's a square?*

Student: *It's in a tilt. But it's a square because if you turned it this way* [motioning a turn with hand], *it would be a square.*

Teacher: *OK. So on your paper, where it asks how you know, you could write exactly what you said, "It's in a tilt. But it's a square because if you turn it, it would be a square."*

In this episode, the teacher's question was critical in determining the student's CBA level of reasoning. Instead of analyzing the properties of the given shape (which would be Level 2 reasoning), this student used holistic reasoning (which is Level 1 reasoning).

Listed below are some questions that can be helpful in conducting individual interviews, interacting with students during small-group work, or conducting a classroom discussion of an assessment task:

- That's interesting; tell me what you did.
- Tell me how you found your answer.
- How did you figure out this problem?
- I'd really like to understand how you're thinking; can you tell me more about it?
- Why did you do that?
- What were you thinking when you moved these objects?
- Did you check your answer to see whether it is correct? How?
- Explain your drawing to me.
- What do these marks that you made mean?
- What were you thinking when you did this part of the problem?
- What do you mean when you say . . . ?

Monitoring the Development of Students' Reasoning

The CBA materials are designed to help you assess levels of reasoning, not levels of students. Indeed, a student might use different levels of reasoning on different tasks. For instance, a student might operate at a higher level when using physical materials than when she does not have physical materials to support her thinking. Also, a student might operate at different levels with shapes that are familiar to her, or that she has done many activities with, as opposed to shapes that are unfamiliar to her. So, rather than attempting to assign a single level to a student, you should analyze a student's reasoning on several assessment tasks, then develop an overall profile of how she is reasoning about the topic. You can see an example of how this is done in Chapter 2.

To carefully monitor and even report to parents the development of student reasoning about particular mathematical topics, many teachers keep detailed records of students' CBA reasoning levels during the school year. To do this, choose several

CBA assessment tasks for each major mathematical topic you will cover during the year. Administer these tasks to all of your students either as individual interviews or as written work at several different times during the school year (for example, before and after each curriculum unit dealing with the topic). In addition to noting the tasks used and the date, record what levels each student used on the tasks.

Differentiating Instruction to Meet Individual Students' Learning Needs

You can tailor instruction to meet individual students' learning needs in several ways.

Individualized Instruction

The most effective way is to work with students individually, using the levels and tasks to precisely assess and guide students' learning. This approach is an extremely powerful way to maximize an individual student's learning.

Instruction by CBA Groups

Another effective way of meeting students' needs is to put students into groups based on their CBA levels of reasoning about a mathematical topic. You can then look to the instructional suggestions for tasks that will be maximally effective for helping the students in each group. For instance, you might have three or four groups in your class, each consisting of students who are reasoning at about the same CBA levels and need the same type of instruction.

Whole-Class Instruction

Another approach that many teachers have used successfully is selecting sets of tasks that all students in a class can benefit from doing. You do this by first determining the different levels of reasoning among students in the class. Then, as you consider possible instructional tasks, ask yourself these questions:

- "How will students at each level of reasoning attempt to do this task?"
- "Can students at different levels of reasoning *succeed* on the task by using different strategies?" (Avoid tasks that some students will not have any way of completing successfully.)
- "How will students at each level benefit by doing the task?"
- "Will seeing how different students do the task help other students progress to higher levels of thinking because they are ready to hear new ways of reasoning about the task?"

Also, sets of tasks can be sequenced so that initial problems target students using lower levels of reasoning while later tasks target students using higher levels.

Another way to individualize whole-class instruction is to ask different questions to students at different levels as you circulate among students working in small groups. For instance, for students who are reasoning about shapes as wholes, you might ask, Is there anything special about the sides and angles in rectangles? Or, if students say that the sides in rectangles are not slanted, you might ask, What do you mean by "not slanted"? How is slant connected to measurements of the angles? Knowledge of CBA levels is invaluable in devising good questions and in asking appropriate questions for different students. In fact, when preparing to teach a lesson, many teachers use levels-of-sophistication descriptions to think about the kinds of questions they will ask students who are functioning at different levels.

Choosing which students to put into small groups for whole-class inquiry-based instruction is also important. If you think of your students' CBA levels of reasoning on a particular type of task as being divided into three groups, you might put students in the high and middle groups together, or students in the middle and low groups together. Generally, putting students in the high and low groups together is not effective because their thinking is likely to be too different.

Assessment and Accountability

As a consequence of state and federal testing and accountability initiatives, most school districts and teachers are looking for materials and methods that will help them achieve state performance benchmarks. CBA is a powerful tool that can help you help your students achieve these benchmarks by:

- monitoring students' development of reasoning about core mathematical ideas;
- identifying students who are having difficulties learning these ideas and diagnosing the nature of these difficulties;
- understanding the nature of weaknesses identified by annual state mathematics assessments results *along with causes for these weaknesses*; and
- understanding a framework for remediating student difficulties in conceptually and cognitively sound ways.

Moving Beyond Deficit Models

The CBA materials can help you move beyond the "deficit" model of traditional diagnosis and remediation. In the deficit model, teachers wait until students fail before attempting to diagnose and remediate their learning problems. CBA offers a more powerful, preventative model for helping students. By using CBA materials to appropriately pretest students on core ideas that are needed for upcoming instructional units, you can identify which students need help and the nature of the help they need before they fail. By then using appropriate instructional activities, you can help students acquire the core knowledge needed to be successful in the upcoming units—making that instruction effective rather than ineffective for these students.

The Research Base

Not only have these materials gone through extensive field testing with both students and teachers, the CBA approach is consistent with major scientific theories describing how students learn mathematics *with understanding*. These theories agree that mathematical ideas must be personally constructed by students as they intentionally try to make sense of situations, and that to be effective, mathematics teaching must carefully guide and support students' construction of personally meaningful mathematical ideas (Baroody and Ginsburg, 1990; Battista, 1999, 2001a; Bransford, Brown, and Cocking, 1999; De Corte, Greer, and Verschaffel, 1996; Greeno, Collins, and Resnick, 1996; Hiebert and Carpenter, 1992; Lester, 1994; National Research Council, 1989; Prawat, 1999; Romberg, 1992; Schoenfeld, 1994; Steffe and Kieren, 1994; von Glasersfeld, 1995). Research shows that when students' current ideas and beliefs are ignored, their development of mathematical understanding suffers. And conversely, "There is a good deal of evidence that learning is enhanced when teachers pay attention to the knowledge and beliefs that learners bring to a learning task, use this knowledge as a starting point for new instruction, and monitor students' changing conceptions as instruction proceeds" (Bransford et al., 1999, p. 11).

The CBA approach is also consistent with research on mathematics teaching. For instance, based on their research in the Cognitively Guided Instruction program, Carpenter and Fennema (1991) concluded that teachers must "have an understanding of the general stages that students pass through in acquiring the concepts and procedures in the domain, the processes that are used to solve different problems at each stage, and the nature of the knowledge that underlies these processes" (p. 11). Indeed, a number of studies have shown that when teachers learn about such research on students' mathematical thinking, they can use that knowledge in ways that positively influence their students' mathematics learning (Carpenter et al., 1998; Cobb et al., 1991; Fennema and Franke, 1992; Fennema et al., 1996; Steffe and D'Ambrosio, 1995). These materials will enable you to:

- develop a detailed understanding of your students' current reasoning about specific mathematical topics, and
- choose learning goals and instructional activities to help your students build on their current ways of reasoning.

Indeed, these materials provide the kind of coherent, detailed, and well-organized research-based knowledge about students' mathematical thinking that research has indicated is important for teaching (Fennema and Franke, 1992).

Research also shows that using formative assessment can produce significant learning gains in all students (Black and Wiliam, 1998). Furthermore, formative assessment can be especially helpful for struggling students, and it can reduce achievement gaps in mathematics learning. The CBA materials offer teachers a powerful type of *formative assessment* that monitors students' learning in ways that enable teaching to be adapted to meet students' learning needs. "For assessment to function formatively, the results have to be used to adjust teaching and learning"

(Black and Wiliam, 1998, p. 142). To implement high-quality formative assessment, the major question that must be asked is, "Do I really know enough about the understanding of my pupils to be able to help each of them?" (Black and Wiliam, 1998, p. 143). CBA materials help answer this question.

Finally, the levels of reasoning described in Chapter 2 are based on a large body of research on students' geometry learning, starting with the original van Hiele theory, which research indicates is accurate in describing the development of students' geometric reasoning (Battista 2007a; Burger and Shaughnessy 1986; Clements and Battista 1992; Fuys, Geddes, and Tischler 1988). However, the CBA levels represent a significant elaboration and clarification of the original van Hiele theory (Battista, 2007a, 2009; Borrow, 2000). Also, research strongly suggests that use of dynamic geometry instruction of the type described in Chapter 3 is quite effective in moving students through the CBA levels (Battista, 2001b, 2001c, 2002a, 2003, 2007a, 2007b, 2008a; Borrow, 2000; de Villiers, 2003; Jones, 2000).

Using CBA Materials for RTI

Response to Intervention (RTI) is a school-based, tiered prevention and intervention model for helping all students learn mathematics. Tier 1 focuses on high-quality classroom instruction for all students. Tier 2 focuses on supplemental, differentiated instruction to address particular needs of students within the classroom context. Tier 3 focuses on intensive individualized instruction for students who are not making adequate progress in Tiers 1 and 2.

CBA can be effectively used for all three RTI tiers. For Tier 1, CBA materials provide extensive, research-based descriptions of the development of students' learning of particular mathematical topics. Research shows that teachers who understand such information about student learning teach in ways that produce greater student achievement. For Tier 2, CBA descriptions enable you to better understand and monitor each student's mathematics learning through observation, embedded assessment, questioning, informal assessment during small-group work, and formal assessment. You can then choose instructional activities that meet your students' learning needs—whole-class tasks that benefit students at all levels; different tasks for small groups of students at the same levels; individualized supplementary student work. For Tier 3, CBA assessments and level-specific instructional suggestions provide road maps and directions for giving struggling students the long-term individualized instruction sequences they need.

Supporting Students' Development of Mathematical Reasoning

CBA materials are designed to help students move to higher levels of reasoning. It is important, however, that instruction not *demand* that students "move up" the levels with insufficient cognitive support. Such demands result in students rotely memorizing procedures that they cannot make personal sense of. *Jumps in levels are made internally by students, not by teachers or the curriculum.* This does not mean that students must progress through the levels without help. Teaching helps students by providing the right kinds of encouragement, support, and challenges—having students work on problems that stretch but do not overwhelm their reasoning, asking good questions, having them discuss their ideas with other students, and sometimes showing them ideas that they don't invent themselves. But when we show students ideas, we should not demand that they use them. Instead, we should try to get students to adopt new ideas because they make personal sense of the ideas and see the new ideas as better than the ideas they currently possess.

Chapter 1

Introduction to Reasoning About Geometric Shapes

Geometry is the mathematical study of spatial objects, relationships, and movements. It uses defined concepts and axiomatic systems to represent and analyze spatial entities. Using geometry to understand the spatial world is foundational for reasoning in mathematics, science, and engineering.

An Overview of How Students Learn Geometry

According to the van Hiele theory, students progress through qualitatively different *levels* of geometric thinking. Research generally indicates that the van Hiele theory is accurate in describing the development of students' geometric reasoning (Battista, 2007a; Burger and Shaughnessy, 1986; Clements and Battista, 1992; Fuys, Geddes, and Tischler, 1988). However, the original levels described by van Hiele contain insufficient detail for CBA purposes. In a series of studies and CBA research, I have elaborated these levels to provide a more detailed account of students' geometry learning (Battista, 2007a).

Level 1: Visual-Holistic Reasoning

Students reason about shapes according to their appearance as visual wholes. They justify their shape identifications using vague holistic judgments such as saying that two figures "look the same." Students often refer to visual prototypes, for example, saying that a figure is a rectangle because "it looks like a door."

Level 2: Analytic-Componential[1] Reasoning

In Level 2, students explicitly attend to, conceptualize, and specify shapes by describing their parts and the spatial relationships between the parts. However, students' descriptions and conceptualizations vary greatly in sophistication. At first, students describe parts and properties of shapes informally and imprecisely using strictly informal language from everyday conversations. As students begin to acquire formal geometric concepts explicitly taught in mathematics curricula (such as angle measure and parallelism), they start to use a combination of informal and formal descriptions of shapes. However, the formal portions of students' shape descriptions are insufficient to completely specify shapes. Finally, students explicitly and exclusively use formal geometric concepts and language to describe and conceptualize shapes in a way that attends to a sufficient set of properties to specify the shapes.

Level 3: Relational-Inferential Property-Based Reasoning[2]

Students explicitly interrelate geometric properties of shapes. For example, a student might say, "If a shape has property X, it also has property Y." However, the sophistication of students' inferences varies greatly. Students start with *empirical* inference, for instance, noticing that whenever they have seen property X occur, property Y occurs. Next, students conclude that when one property occurs, another property must occur by *analyzing how shapes can be built* one part at a time. For instance, students might conclude that if a quadrilateral has four right angles its opposite sides are equal because when they draw a rectangle by making a sequence of perpendiculars they cannot make the opposite sides unequal.

Next, students make simple *logical* inferences about properties. For example, a student might reason that because a square has all sides equal, it has opposite sides equal. Such reasoning enables students to understand hierarchical classification of shapes. For instance, a student whose definition for a rectangle is "4 right angles and opposite sides equal" might infer that a square is a rectangle because "a square has 4 right angles, which a rectangle has to have; and because a square has 4 equal sides, it has opposite sides equal, which a rectangle has to have." In the final phase of Level 3, students use logical inference *to reorganize* their classification of shapes into a logical hierarchy. It not only becomes clear why a square is a rectangle, but classifying a square as a rectangle becomes a necessary part of reasoning. Throughout Level 3, students are increasingly able to understand and appreciate minimal definitions for classes of shapes; that is, definitions that list only enough properties to specify the class of shapes, not all the properties that the class has.

Level 4: Formal Deductive Proof

Students understand and can construct formal geometric proofs. That is, within an axiomatic system, they can produce a sequence of statements that logically justifies

---
[1] I use the term *analysis* to refer to the process of understanding objects by decomposing them into components.
[2] The description for Level 3 was developed by Borrow (2000) and Battista.

a conclusion as a consequence of the "givens." They recognize differences among undefined terms, definitions, axioms, and theorems.

Mental Processes in Geometric Reasoning and Learning

Abstraction is the process by which the mind selects, coordinates, unifies, and registers in memory a collection of mental objects or acts. Two special forms of abstraction are fundamental to geometry learning and reasoning (Battista, 2002b). *Spatial structuring* is the act of creating an organization or form for an object or set of objects. It determines an object's nature or shape by identifying its spatial components and establishing interrelationships between those components. *Mental models* are nonverbal image-like[3] mental versions of situations that have structures isomorphic to the perceived structures of the situations they represent (Greeno, 1991; Johnson-Laird, 1983, 1998). Individuals reason about situations by activating mental models that enable them to simulate interactions within the situations so that they can explore possible scenarios and solutions to problems.

Geometry as the Study of Structure

One way of thinking about geometry is as the study of ways of organizing or *structuring* our spatial environment (Battista, 2007a). When we structure something, we determine its nature, shape, or organization by decomposing it into parts and establishing interrelationships between these parts. We structure the plane and space when we organize them by coordinate systems. We structure our visual world when we view it in terms of shapes such as lines, angles, polygons, polyhedra, and geometric transformations.

Spatial Structuring

The most fundamental process in geometric thinking is spatial structuring. Spatial structuring determines how an individual perceives an object's shape by identifying its components, combining components into composites, and establishing interrelationships between and among components and composites (such as how they are placed in relation to each other). It results in mental models of environments and objects that enable us to move about in them, manipulate or operate on them, and reflect on and analyze them.

Geometric Structuring

Geometric structuring uses formal geometric concepts to represent or analyze a spatial structuring. Geometric structuring involves using specific formal concepts

---

[3] Images do not have to be visual; they also can be kinesthetic (or even aural). For instance, you can have the image/feeling of what it is like to throw a baseball.

such as angles, slope, parallelism, length, rectangle, coordinate systems, and geometric transformations to conceptualize and operate on spatial situations. For a geometric structuring to make sense to a student, its description must create or be connected to an appropriate spatial structuring.

Axiomatic Structuring

Axiomatic structuring formally organizes geometric concepts into a system and specifies that interrelationships must be described and established through logical deductions.

Objects of Geometric Analysis

Students reason about three distinct types of "objects" in geometry (Battista, 2007a, 2007b). *Physical objects* are things such as a door, box, ball, geoboard figure, picture, drawing, or dynamic computer figure. *Concepts* or conceptualizations are the mental representations that individuals abstract for categories of like objects. *Concept definitions* are verbal or symbolic specifications of categories of objects or entities. Importantly, students' concepts can be very different from what formal concept definitions specify. Also, during instruction students and teachers often focus on different objects.

For example, diagrams and physical objects play two major roles in geometry. On one hand, physical objects can be thought of as the *input* for geometric conceptualization. Geometric concepts are generally created through an analysis of such objects: the objects become *represented* or described by the concepts. On the other hand, physical objects, including diagrams, are often used to *represent* formal geometric concepts. As an example, we can think of a triangle diagram as representing the formal concept of triangles, or as a single object that we wish to analyze geometrically. Students often get confused about the represented and representing role of diagrams. That is, when teachers use diagrams to represent formal concepts, students often reason about the diagrams, not the formal concepts. Consequently, students often attribute irrelevant characteristics of a diagram to the geometric concept it is intended to represent (Clements and Battista, 1992; Yerushalmy and Chazan, 1993). For instance, students might not recognize right triangles in nonstandard orientations because they are thinking about prototypical right triangles with legs that are vertical and horizontal instead of the formal abstract concept of right triangle. They are reasoning about specific instances rather than the abstract idea of right triangle given by a formal definition.

Two Kinds of Concepts

Another issue in geometry learning is that people learn two different kinds of concepts (Pinker, 1997). *Natural concepts*, such as apples or dogs, are formed in everyday activity and are rarely accompanied by concept definitions. Such concepts generally are induced from instances and are thought about in terms of visual resemblances to prototypical examples. *Formal concepts*, such as rectangles, have definitions that explicitly specify a sufficient set of properties to identify instances.

As students try to learn geometry, they almost always form both natural and formal concepts. Before students are taught geometry, and even in primary-grade teaching, students learn to identify but not define shapes—natural shape concepts are formed. Subsequent school instruction demands that students learn formal concepts for shapes in terms of verbally stated, property-based definitions. Difficulties often arise as students attempt to reconcile natural and formal concepts for the same category. The names used for the natural and formal concepts may be the same, but the underlying cognitive entities are vastly different. For instance, students who possess a natural concept of square think of it as a particular kind of image, whereas students with a formal concept think of it as a shape that possesses a specific set of properties. As we will see, moving from CBA Level 1 to 2 involves progressing from natural to formal shape concepts.

Especially important in this development are formal concepts that are used to specify relationships between parts of shapes. The concepts of angle measure, length measure, congruence, and parallelism are used to specify the defining spatial properties of shapes. For example, the statement "Opposite sides of a rectangle are the same length" uses the concept of length to describe the relationship that exists between the sides of rectangles. The statement "A rectangle has four right angles" uses the concept of right angle to describe a relationship between pairs of adjacent sides (angle measure tells how much one side of an angle must be rotated to coincide with the other side). The statement "This quadrilateral has one line of symmetry" uses the concept of symmetry to describe a relationship between two "halves" of a quadrilateral. So defining a square as a four-sided polygon that has four 90 degree angles and all sides the same length specifies the spatial properties that characterize the category of shapes we call squares.

In summary, understanding geometric properties is core knowledge in geometry (Battista, 2007a, 2007b, 2008a, 2009). Without it, students' reasoning about shapes is informal, imprecise, and usually incorrect. To understand and guide students' development of geometric reasoning, we must trace students' progress in moving from informal, vague conceptualizations of shapes to the formal, property-based conceptual system used by mathematicians. This chapter and the next describe the levels of sophistication in students' development of this reasoning.

Understanding Students' Levels of Sophistication in Geometric Reasoning

The CBA levels provide a detailed description of the development of students' reasoning about geometric shapes. This detail is critical for tailoring instruction to meet students' learning needs. However, when you are first learning a set of CBA levels, the amount of detail can be overwhelming. So, keep in mind that understanding CBA levels comes in stages and develops over time. First, you will learn the major features of the levels-of-sophistication framework for geometric shapes. As you use CBA with your students, you will learn the details of the framework.

Zooming Out to Get an Overview

To begin understanding the CBA levels for geometric shapes, it is important to develop an understanding of the overall organization of the levels. The chart below shows the CBA geometric shapes levels in a "zoomed out" view that corresponds to the original van Hiele levels.

CBA Levels of Sophistication in Students' Reasoning About Geometric Shapes	
Level	Description
1	Student identifies shapes as visual wholes.
2	Student describes parts and properties of shapes.
3	Student interrelates properties and categories of shapes.
4	Student understands and creates formal deductive proofs.

Zooming in to Meet Individual Students' Needs

Understanding individual students' reasoning precisely enough to maximize their learning or remediate a learning difficulty requires a detailed picture of that learning. The "jumps" between levels must be small enough that students can achieve them with small amounts of instruction in relatively short periods of time.

Imagine students trying to climb to a cognitive plateau needed to meet an instructional goal. In situation A, the student has to make a cognitive jump that is too

Figure 1.1 Accessible Cognitive Jumps

great. In situation B, the student can get to the goal by using accessible CBA levels as stepping-stones. To provide students the instructional guidance and cognitive support they need to develop a thorough understanding of mathematical ideas, you need to understand these sublevels, which correspond to the elaboration provided by Battista (2007a). Chapter 2 provides detailed descriptions and illustrations of all the CBA levels and sublevels for geometric shapes.

Components of Sophistication: Type and Validity of Reasoning

Two major components must be considered in analyzing the sophistication of students' geometric reasoning. The first component is *type of reasoning*, which has several levels and ranges from visual-holistic reasoning to formal proof. The second component is *validity of reasoning*, which involves the accuracy and precision of students' identifications, descriptions, conceptualizations, explanations, and justifications. Validity is judged relative to level and type of reasoning. For instance, the validity of Level 1 visual-holistic reasoning is judged by accuracy in identifying shapes. In contrast, the validity of Level 4 reasoning is judged by correctness of formal proofs. This book focuses on types of reasoning, with the validity component interwoven throughout the discussion.

Different Levels for Different Shapes

What, precisely, does it mean to be "at" a CBA level? Essentially, students are at a level when their overall cognitive structures and processing causes them to be disposed to and capable of thinking about a topic in a particular way. So students are at CBA Level 1 when their overall cognitive organization and processing disposes them to think about geometric shapes in terms of visual wholes; they are at Level 2 when their overall cognitive organization disposes and enables them to think about shapes in terms of their properties. Also, when students move from familiar content to unfamiliar content, their level of thinking might decrease temporarily; but because students are disposed to operate at the higher level, they attempt to use that level on the new material and quickly become capable of using that level (Battista, 2007a). For instance, in moving from studying quadrilaterals to studying triangles, students who are at Level 2 for quadrilaterals might initially process triangles as visual wholes, but right from the start they look for, and fairly quickly discover and use, triangle properties. Finally, when transitioning between two levels, students might use different reasoning on different tasks. For instance, a student might use Level 2.3 reasoning to define a rectangle (which is more familiar) as a quadrilateral with four right angles and opposite sides equal, but use Level 2.1 reasoning to define a parallelogram (which is less familiar) as a quadrilateral that has opposite sides "even."

Chapter 2

Levels of Sophistication in Student Reasoning: Geometric Shapes

The CBA approach to guiding students' development of understanding of length builds on the CBA levels of sophistication in students' reasoning. Understanding these levels enables teachers to tailor instruction to meet students' individual learning needs. The major CBA levels (Levels 1–4) provide an overview of the ways students think about geometric shapes.

Understanding students' reasoning precisely enough to maximize their learning or remediate their learning difficulties requires a more detailed picture than is provided by the major levels, so the major levels are divided into sublevels. The "jumps" between sublevels are small enough that students can achieve them with small amounts of instruction in relatively short periods of time. Sublevels serve as accessible stepping-stones in students' development.

The following chart summarizes the CBA levels for understanding of geometric shapes. The following pages provide a detailed description of each level, along with examples of student work at each level. At first glance, the amount of detail in the CBA levels can be overwhelming. So keep in mind that understanding CBA levels develops as you study examples of students' work and as you use CBA with your students.

| \multicolumn{4}{c}{Geometric Shapes} |
Level	Sublevel	Description	Page
1		**Student identifies shapes as visual wholes.**	9
	1.1	Student incorrectly identifies shapes as visual wholes.	10
	1.2	Student correctly identifies shapes as visual wholes.	15
2		**Student describes parts and properties of shapes.**	19
	2.1	Student informally describes parts and properties of shapes.	20
	2.2	Student uses informal and insufficient formal descriptions of shapes.	27
	2.3	Student formally describes shapes completely and correctly.	33
3		**Student interrelates properties and categories of shapes.**	37
	3.1	Student uses empirical evidence to interrelate properties and categories of shapes.	37
	3.2	Student analyzes shape construction to interrelate properties and categories of shapes.	38
	3.3	Student uses logical inference to relate properties and understand minimal definitions.	40
	3.4	Student understands and adopts hierarchical classifications of shape classes.	43
4		**Student understands and creates formal deductive proofs.**	45

LEVEL 1: Student Identifies Shapes as Visual Wholes

Students identify, describe, and reason about shapes according to their appearance as visual wholes. They often compare shapes to well-known examples, saying, for instance, that a shape is a rectangle because "it looks like a door." Or they might decide that two shapes are the same simply because they "look the same." Students describe shapes holistically rather than by their parts. For instance, in distinguishing rectangles from squares, students might say that rectangles are "long" and squares are "fat." As another example, they might say that triangles are "pointy" and circles are "round."

Orientation of figures strongly affects students' reasoning. For example, students often correctly identify a square only if its sides are horizontal and vertical. Students often justify their shape identifications using imagined visual transformations, saying, for instance, that a shape is a square because "if it is turned it would be a square."

Although students who are reasoning at Level 1 correctly identify some familiar geometric shapes, they also incorrectly identify many shapes. For instance, such students may not recognize atypical triangles as triangles, or they may not distinguish rhombuses from parallelograms that do not have all sides equal. Students have difficulty recognizing commonalities in groups of shapes that share a set of properties.

Level 1.1 Student incorrectly identifies shapes as visual wholes.

Students' inaccurate visual-holistic reasoning prevents them from identifying many common shapes. One reason for this inaccuracy is that students at this level do not construct and operate on accurate images of shapes—their imagery distorts shapes in critically important ways.

EXAMPLES

Task: *Tell what kinds of shapes are shown below.*

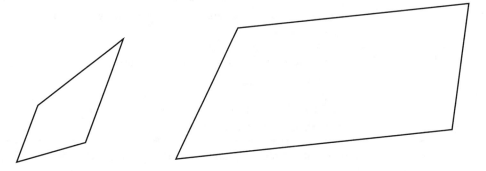

Response: That kinda looks like a kite.

Response: It's a parallelogram.

Task: *Tell what kinds of shapes are shown below.*

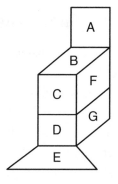

Response: I think *[Shape]* A is a square and *[Shape]* F is a rectangle.

Task: *Tell what kind of shape is shown below.*

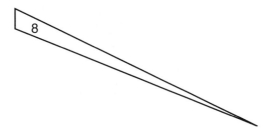

Response: *[Rotates paper several times]* It just doesn't look like a triangle. It is just like a point.

Task: *How are these shapes alike?*

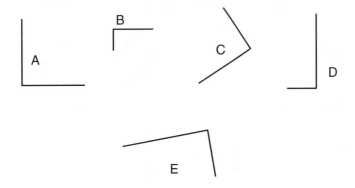

Response: I think what is alike about them is that they all kind of bend. None of them really zigzag. *[Teacher: When you say zigzag, what do you mean? Can you show me by drawing?] [Student draws the following shape.]* Yeah, none of them are like that. They are all bent.

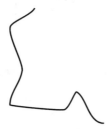

This student comments about figures as wholes. She does not address, even informally, that the shapes have right angles.

Levels of Sophistication in Student Reasoning: Geometric Shapes

Task: *What kind of shape is this [the shape below is not a square]?*

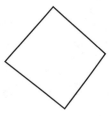

Response: It's a square. *[Teacher: How do you know it's a square?]* It's in a tilt. But it's a square because if you turned it this way *[motioning with her hand]*, it would be a square.

Task: *A special drawing machine makes shapes like this [see A–E below]. All the shapes made by this drawing machine are alike in some way.*

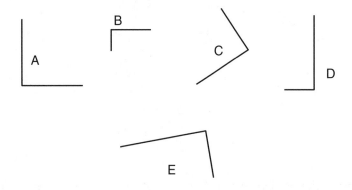

Shape F cannot be made by this drawing machine.

Which of these shapes [G–I] can be made by the drawing machine?

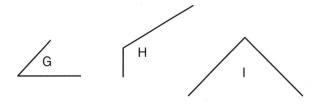

Response: *[Circles Shape G]* I think this one can be made by the drawing machine. *[Teacher: Why?]* Because this one isn't diagonalled. It is straight just like the others. And it is curved. And it looks just like this one *[pointing to Shape C]*.

Task: *Circle the rectangles.*

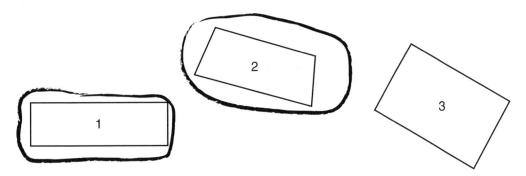

Response: *[Circles shapes as shown] [Teacher: Why did you say that Shape 2 is a rectangle?]* It's just like that one *[Shape 1]* only it's crooked. *[Teacher: Why did you say that Shape 3 is not a rectangle?]* It's just like this one *[Shape 2]* only it's way too fat, and it is too, like, slanted.

Task: *[Show students the set of shapes below, each of which has all sides equal.] All the shapes made by this drawing machine are alike in some way. Can you figure out what kind of shapes this drawing machine makes? Look at the shapes the drawing machine makes. Describe a rule for making these shapes. What's alike for all of them?*

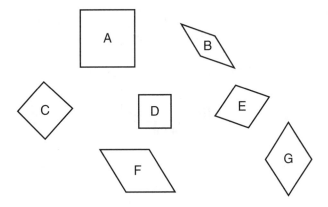

Response: There's 3 different shapes here.

[GROUP 1: Shapes A, D, and E] It's *[pointing at A]* the same as this *[pointing at D]*. Shape E is like Shape D because you could just flip it *[E]* over like this, so they would be the same.

[GROUP 2: Shapes C and G] This *[pointing at C]* is exactly like this *[pointing at G]* only it's wider. This *[G]* is longer.

[GROUP 3: Shapes B and F] These are the same but one *[F]* is bigger and wider.

[Teacher: So what's the rule for these shapes?] The rule is that it makes them larger and fatter.

Typical of students at this level, the student was unable to see the common characteristic of all the shapes shown above (that each one is a rhombus and thus has all 4 sides the same length).

Task: *Explain why you think Shapes 1 and 9 are squares.*

Response: Number 9 is a square because it's not long like a rectangle, and it's fat like Number 1. It's big, but it's not long.

Task: *Describe exactly how you decide if a shape is a rectangle or not.*

Response: A square is shorter than a rectangle, so rectangles will be longer, so that's how I know it will be a rectangle or a square.

Task: *Tell whether this shape is a rectangle.*

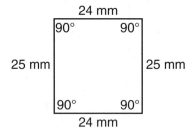

Response: No, because it is not long.

Task: *Circle each rectangle. Then describe everything you know about rectangles.*

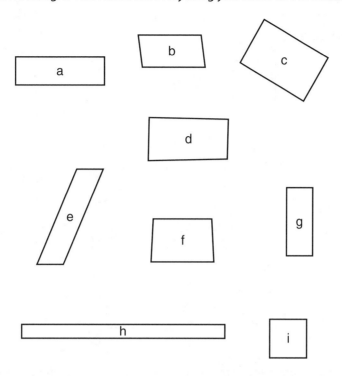

Response: *[Circles Shapes a, b, d, e, g, and h]* Rectangles stand up straight or they lay down.

For strategies to help students at Level 1.1, see Chapter 3, page 70.

Level 1.2 Student correctly identifies shapes as visual wholes.

Students use visual-holistic reasoning to *correctly* identify familiar geometric shapes. They often recognize when a group of shapes possesses a common set of geometric properties (although they cannot accurately describe these properties). Students at this level can form accurate visual images of geometric figures, and they can manipulate these images in ways that maintain the images' structure and identity.

EXAMPLES

Task: *Circle each square.*

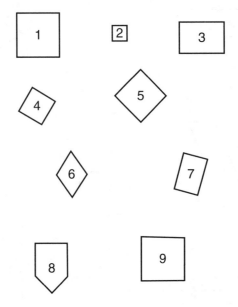

Response: *[After correctly circling the squares, student explains why Shapes 3, 7, and 8 are NOT squares.]* Shape 3 is not a square because usually a square is not stretched out and this has more added on to it. Shape 7 is longer than a square would be. If Shape 8 was a square, it wouldn't have this little bit *[bottom 2 sides]* added.

Task: *Why did you say that this shape is a triangle?*

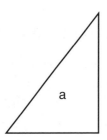

Response: Because if you turn it around, it will sort of look like a triangle *[turning the page 90° clockwise]*.

Task: A special drawing machine made these shapes *[showing Shapes A–G]. How are these shapes alike? What's the rule?*

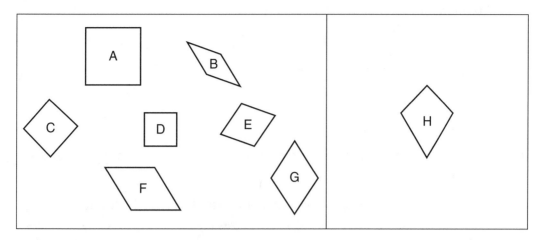

Response: They all kind of look like diamonds. *[Teacher: How do you know those are diamonds?]* Because if you turned them, they would look like diamonds. *[Teacher: Shape H cannot be made by the drawing machine. Why?]* It looks like a kite, but no one else looks like a kite.

Even though "diamond" is an imprecise, holistic description of the example shapes, the student's conceptualization is sufficient to correctly identify examples and nonexamples of shapes like A–G.

Task: *Drawing Machine 1 makes shapes like this [A–E]. All the shapes made by Drawing Machine 1 are alike in some way. They follow a rule. Your job is to figure out what kind of shapes Drawing Machine 1 makes. How are the shapes made by this drawing machine alike?*

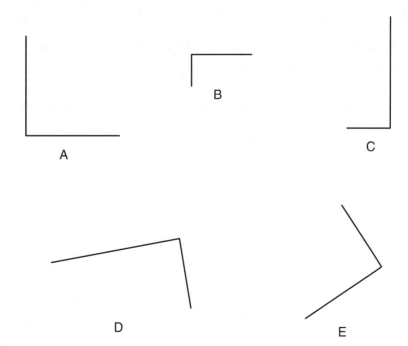

Levels of Sophistication in Student Reasoning: Geometric Shapes

Response: They are all shaped like L's. *[Teacher: If you had to write a rule, what would your rule say?]* An L-making machine. *[Teacher: Which of these shapes can be made by Drawing Machine 1?]*

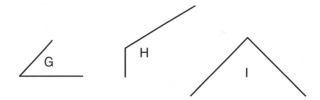

[Loops Angle I] Because if you look at it this way *[rotates paper]*, it looks like an L and if you look at these two *[H and G]* they're diagonal.

Although this student reasons about shapes holistically, he properly discriminates examples from nonexamples. He visually recognizes that Drawing Machine 1 makes right angles.

Task: *Which shapes are rectangles?*

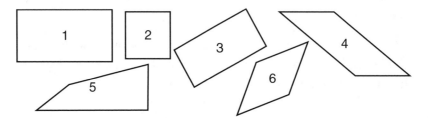

Response: Shape 1 is a rectangle. Shape 2 is a rectangle. And so is Shape 3—it's just slanted up. Shapes 5, 6, and 4 aren't rectangles. *[Teacher: Why aren't they rectangles?]* I don't know why they're not. This one *[points to Shape 4]* could be a rectangle if it were slanted upward. These 2 *[points to 5 and 6]* aren't even close to being a rectangle.

Although this student accurately identifies rectangles, his explanations for the identifications are vague and holistic.

Task: *Shapes A–G can be made by a special drawing machine. Shape H cannot.*

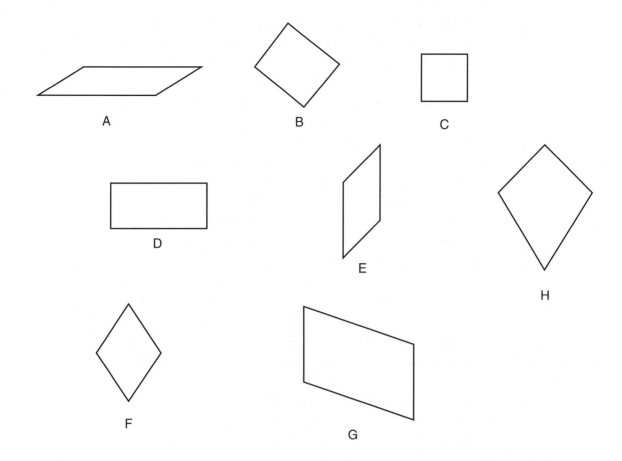

Which of these shapes can be made by this drawing machine? Why or why not?

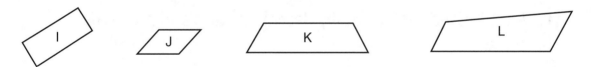

Response: *[Correctly circles Shape I, then turns the paper and circles Shape J. He turns the paper for the other two shapes but does not circle them.]* [Teacher: Why do you think those two shapes (I and J) *can be made?]* Because you can turn them around and they will still look the same. *[Teacher: How about the other two* (Shapes K and L)*?]* If you turn them upside down they look different.

For strategies to help students at Level 1.2, see Chapter 3, page 73.

LEVEL 2: Student Describes Parts and Properties of Shapes

Students recognize, specify, and analyze shapes by identifying the shapes' parts and properties. The *properties* of a shape are those spatial relationships between its parts that characterize the shape.

Levels of Sophistication in Student Reasoning: Geometric Shapes

At Level 2, students' descriptions and conceptualizations first focus on specifying parts of shapes, then on describing spatial relationships between parts. Examples of identifying *parts of a shape* are counting its sides or angles (usually referred to as "points") or saying that it has "straight" sides (meaning sides with no bending or turning). Examples of describing *relationships between parts of a shape* are saying that all of its side lengths must be equal or that all of its angles must measure 90° (specifying angle measure tells how adjacent sides are related).

Students' descriptions and conceptualizations of shapes' parts and properties vary greatly in sophistication. Initially, students' descriptions are visual, informal, and imprecise (Sublevel 2.1). At the end of Level 2, students' descriptions are complete and correctly use formal geometric terms (Sublevel 2.3). (Formal geometric terms are those that are explicitly taught in mathematics curricula, such as length of sides, congruence, degree measure of angles, right angles, parallelism, or perpendicularity.)

Two major interrelated factors contribute to the development of Level 2 reasoning. The first is an increasing ability and inclination to account for the spatial structure of shapes by analyzing their parts and how the parts are interrelated. The second is an increasing ability to understand and use formal geometric concepts (such as length, angle measure, and parallelism) in specifying relationships between parts of shapes.

Throughout Level 2, students define a class of shapes by listing all the properties they know about the shape. For instance, they might define a rectangle as a quadrilateral having four 90° angles, opposite sides equal, and opposite sides parallel. Students do not understand definitions that list only a sufficient property such as "a rectangle is a quadrilateral having four 90° angles" until they reach Level 3.4.

Level 2.1 Student informally describes parts and properties of shapes.

Using language learned in everyday conversation, students describe parts and properties of shapes visually, informally, and imprecisely. For example, as informal references to the property of having all right angles, students might say that rectangles have "square corners" or "straight sides" or "sides that are not at an angle." When students say that rectangles and squares have "square corners," we have some evidence that they have noticed that rectangles and squares have the same special type of angles. But because the term *square corners* is not formally and precisely defined, it is difficult to know exactly what students mean. And when students say that rectangles have "straight sides" or "sides that are not at an angle," they are using the formal terms of *straight* and *angle* in informal ways, presumably as a reference to the perpendicularity of adjacent sides. However, sometimes when students use the term *straight* to describe a rectangle's angles, they mean that the sides of a rectangle are horizontal and vertical. (Think about when a picture hanging on a wall is considered "straight" or "not at an angle.")

It is important to recognize that there is always an inherent imprecision in informal descriptions. For instance, when students characterize rectangles as having

"2 long sides and 2 short sides," they may or may not mean that the opposite sides of a rectangle are the same length. But no matter what they mean, their description of rectangles as having "two long sides and two short sides" applies to many other quadrilaterals (as shown in the diagram below). So this description is too imprecise to accurately characterize one of the essential properties of rectangles.

At Level 2.1, students' descriptions often only make sense when they are accompanied by gestures (for example, students might say that "this is equal to this," which makes sense only if students point to portions of figures to show what each "this" refers to). This need to gesture indicates imprecision in property descriptions (and often reasoning). Also, at Level 2.1, students seem to describe shapes as they inspect particular shapes as opposed to describing properties that they have previously decided all shapes in a group have.

Students often mix informal descriptions of parts and properties of shapes with the holistic-visual reasoning about the shapes they used in Level 1.

EXAMPLES

Task: *A special drawing machine makes shapes like this [see below]. All the shapes made by this drawing machine are alike in some way. They follow a rule. Your job is to figure out what kind of shapes this drawing machine makes. How are the shapes made by this drawing machine alike?*

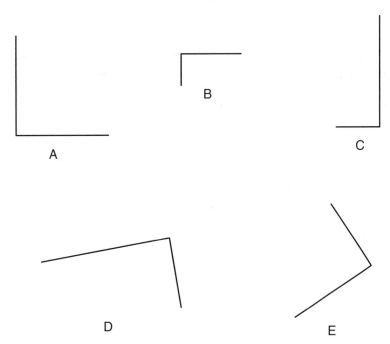

Response 1: Like they all have little corners.

The student's reference to shape parts—corners—is not precise enough to discriminate examples (right angles) from nonexamples (angles that are not right angles).

Response 2: They all got one point *[pointing at the vertices of the angles]*. They all have two strings. *[Teacher: Two?]* Lines. . . . *[Teacher: What is your rule again?]* They all have one point and they all have two lines. *[Teacher: This shape cannot be made by the drawing machine. Why?*

[Turns the student sheet 90°, then 90° again] I am thinking. It has two lines like the other shapes *[A–E]*. But I can't think why it can't be made by the drawing machine.

This student informally describes the components of the shapes ("points," "strings," "lines"). (His use of the term "point" seems to refer to a sharp, object-like portion of a figure, not a geometric point, which is a location in space.)

Task: *Why did you say that this shape is a triangle?*

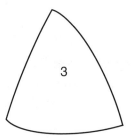

Response: Well, it is kind of sloppy but it looks like one because of the points at the bottom and the point at the top. *[Teacher: Why is it sloppy?]* Because the bottom kind of hangs out *[tracing the bottom of the figure]* and the lines aren't straight.

The student refers to shape parts—vertices and sides—informally as "points" and "lines."

Task: *How are these shapes alike? What's the rule?*

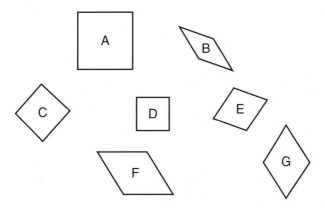

Response: They all have 4 lines or a slant. Not a slant. Four points *[pointing to each of the 4 vertices for a few of the Shapes A–G]. . . . [Teacher: So, what do you think the rule is?]* They all have 4 lines and 4 slants. I mean not 4 slants, 4 points.

This student's informal parts-based description is not only imprecise, it does not deal at all with the defining characteristic of Shapes A–G—all sides are equal in length (so all the shapes are rhombuses).

Task: *All shapes made by this drawing machine are alike in some way. Can you figure out what kind of shapes this drawing machine makes? Describe a rule for making these shapes. What's alike for all the shapes?*

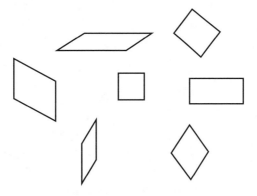

Response 1: They all have 4 corners. That is really it; nothing else really.

Response 2: They all have sharp corners. And they're all made out of straight lines, they don't have squiggly lines; no lines are curved, nothing like that. No matter what, they don't have squigglies; they're all straight *[traces one edge of a shape]. [Teacher: What do you mean when you say straight?]* They're all straight. I mean they're not like curved or squiggly *[draws a curve and a squiggle]*.

Levels of Sophistication in Student Reasoning: Geometric Shapes

Task: *Why did you say that Shape 4 was not a rectangle?*

Response: Because these two are uneven *[pointing to the left and right sides]*; this one is smaller and this one is longer *[first pointing to the right side, then the left side of Shape 4]*. And this one is kind of going up like that, and this one is kind of going straight *[pointing to the top and bottom sides]*.

Although this student's language is informal, her description attends to properties that must be possessed by rectangles. Because she is describing why Shape 4 is *not* a rectangle, she only needs to say that one property necessary for rectangles is not possessed by this shape.

Task: *What do you think the rule is for the shapes made by Drawing Machine 1 [Shapes A–E]?*

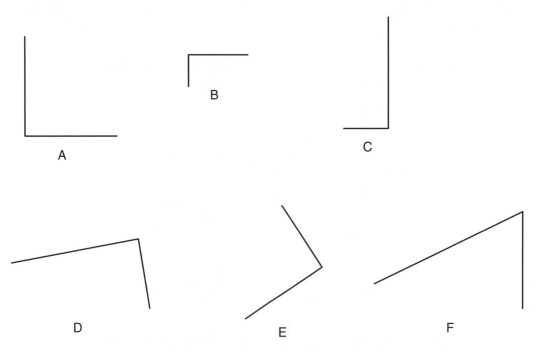

Response: These *[Shapes A–E]* don't have slants. . . . This one has a slant *[turning the paper 90° and pointing to Shape F, which cannot be made by Drawing Machine 1]*. *[Teacher: Which of these shapes can be made by Drawing Machine 1? Why? Why not?]*

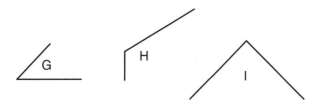

[Student circles only Shape I.] [Teacher: And why do you think that?] Because it doesn't have a slant. Because if you turn it *[turning the paper 90°]*, it doesn't have a slant. If you turn this *[Shape H]*, it has a slant, and this one has a slant *[crossing out Shape G]*. This one has a slant *[crossing out Shape H]*. And this one doesn't have a slant *[Shape I]*.

By "slant," the student appears to mean "not perpendicular," which is the common attribute of Shapes A–E. But the word "slant" is informal and imprecise.

Task: *Which of these shapes can be made by a drawing machine that makes rectangles, that is, a Rectangle Maker?*

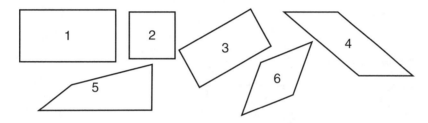

Response: The Rectangle Maker cannot make Shape 4 because the Rectangle Maker has no slants. The Rectangle Maker cannot make Shape 6 because Shape 6's sides are at a slant, and the Rectangle Maker always has lines that aren't at a slant.

Apparently, this student used the informal term *slant* to describe the concept of not being perpendicular, which correctly discriminated rectangles from parallelograms.

Task: *How do you know if a shape is a parallelogram?*

Response: In a parallelogram, the sides across from each other have to be straight or slanted. . . . If this side is slanted, this side has to be slanted also *[pointing to both sets of opposite sides—see below]*. *[Teacher: Let me understand this. So if this side is slanted, this one has to be slanted?]* And if it's straight, the other one has to be straight. . . . *[Teacher: I'm not sure I totally understand.]* It's like the sides across from each other have to be slanted the same way.

Levels of Sophistication in Student Reasoning: Geometric Shapes

Although her description is vague and imprecise, the student is explicitly attending to a necessary spatial relationship between the sides of a parallelogram. In essence, it seems that for this student, two sides are parallel if they are slanted the same way. (Mathematically, two lines in a plane are parallel if they have the same slope, which is the mathematical concept for "slant.")

Task: *Why did you say that this shape is not a square?*

Response: It has like the same length around here and here and here and here *[indicates 4 sides]*. But it's the way the lines are put together; it is not put together like a square would be.

Task: *Is this shape a square?*

Response: No, because it's long, and all the sides aren't the same.

Saying that "the sides aren't the same" is not quite precise enough to be a formal description—the student must say that the *lengths* of the sides are not the same, or that the sides are not *congruent*, or even that the sides are not equal. Also, note that this student related her property-based reasoning (the sides not being the same) to visual-holistic reasoning (it's long). For her, it seems that the informal description, "All sides aren't the same" is an informal property-based description that explains her idea that the shape is "long."

For strategies to help students at Level 2.1, see Chapter 3, page 75.

Level 2.2 Student uses informal and insufficient formal descriptions of shapes.

When describing the parts or properties of shapes, students use a combination of informal and formal descriptions. The formal descriptions utilize standard geometric concepts and terms explicitly taught in mathematics curricula, but they are insufficient to completely specify shapes. For example, a student might explain that a rectangle has "sides across from each other that are equal [formal] and square corners [informal]." In this case, the formal portion of the student's description—that opposite sides are equal—is insufficient to specify rectangles. Students' formal language can be inadequate by being incorrect, insufficient, or inconsistent. Although students often recall properties they have abstracted for classes of shapes (say, "two long sides and two short sides" for rectangles), their reasoning is still visually based, and most of their descriptions and conceptualizations seem to occur extemporaneously as they are inspecting shapes.

Incorrect Descriptions

The formal portions of students' shape descriptions are incorrect.

EXAMPLES

Task: *Why did you say that this shape is not a triangle?*

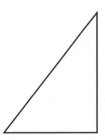

Response: No 2 sides have equal lengths, so it is not a triangle.

This student uses a formal property—equal side lengths—to specify triangles. But the formal property she attributes to triangles is incorrect. Having 2 equal side lengths is a property that specifies isosceles triangles, not all triangles.

Task: *Explain what a hexagon is.*

Response: A hexagon is a 5-sided polygon.

This student likely knows what the formal terms "side" and "polygon" mean. But he does not know how many sides a hexagon has. (A pentagon has 5 sides; a hexagon has 6 sides.)

Task: *Is this shape a rectangle?*

Response: Yes. It is a rectangle because rectangles have 4 sides.

This student's formal specification of "4 sides" is insufficient to specify a rectangle. Note that, in this case, the correct answer arises from incorrect reasoning.

Insufficient Descriptions

The *formal* portions of students' shape descriptions are insufficient to specify the shape. A set of properties is *sufficient* to characterize a type of shape if any shape that has the properties MUST be that type of shape. For instance, a set of properties sufficient to characterize a shape as a rectangle is:

- The shape must be a quadrilateral.
- The shape must have opposite sides equal in length.
- The shape must have opposite sides parallel.
- The shape must have all angles right angles.

Any shape that has these 4 properties must be a rectangle. Another sufficient set of properties for the class of rectangles is:

- The shape must be a quadrilateral.
- The shape must have all angles right angles.

Any shape that has these two properties must be a rectangle. However, understanding that this second set is sufficient requires a student to infer properties from other properties, and does not develop until students reach Level 3 reasoning.

EXAMPLES

Task: *What are all the special things about a rectangle?*

Response 1: A rectangle has sides across from each other that are equal lengths.

Response 2: The sides across from each other are equal *[formal]* and it has square corners *[informal]*.

The first student correctly and formally describes one necessary property for being a rectangle—opposite sides are equal in length. But she does not say anything about another necessary property—that rectangles have 4 right angles. So her formal description does not correctly specify rectangles. (It actually specifies parallelograms.)

The second student also formally and correctly specifies that opposite sides of rectangles are the same length. Furthermore, he says something about the angles of rectangles. But what he says—that they have "square corners"—informally and imprecisely describes right angles. So the formal part of his description is insufficient.

Task: *All shapes [A–G] made by a special drawing machine are alike in some way. Can you figure out what kind of shapes this drawing machine makes? How are these shapes [A–G] alike?*

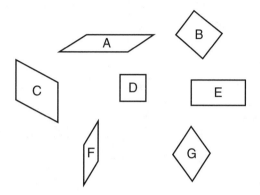

Response: They don't have equal sides. This side is bigger than this side *[marking adjacent sides for several of the shapes, as shown]*.

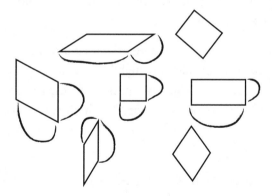

The student uses the formal property "don't have equal sides" to describe shapes made by the drawing machine. But his property inadequately describes what Shapes A–G have in common (they are all parallelograms, so their opposite sides are parallel and congruent). For instance, squares (like Shape D) do not possess this property, and shapes such as that below, which cannot be made by the drawing machine, possess the "don't have equal sides" property.

Levels of Sophistication in Student Reasoning: Geometric Shapes

Task: *After correctly identifying a shape as a rectangle, the student is asked why the shape is a rectangle.*

Response: In rectangles, the sides that are across, like this one and this one, they are always the same length, and this one and this one *[two shorter sides]* are always the same length.

This student uses the formal concept—opposite sides equal in length—to describe rectangles. However, her description is insufficient because she fails to mention that rectangles must have right angles. So her description does not discriminate rectangles from parallelograms.

Task: *Circle each square.*

Response: Usually a square has the same length on each side. *[Shapes]* 1 and 2 have that. They are not stretched out too far so they make rectangles. *[Shape]* 6 has the same length around here *[indicates sides]*. But it's the way the lines are put together; it is not put together like a square would be.

Although this student reasons about side lengths formally and correctly, her reasoning about the angles in squares is informal. She seems to know visually that a square has a special kind of angles, but she cannot express this idea formally by talking about right angles or perpendicularity. So the formal part of her description is insufficient.

Descriptions Inconsistent with Identifications

The formal portions of students' shape descriptions are inconsistent with their identifications of the described shapes. This category includes rote use of formal geometric terms—that is, when students use formal terms without understanding the meaning of those terms.

EXAMPLES

Task: *What are all the special things about a rectangle?*

Response 1: In a rectangle, the sides next to each other have to be perpendicular. *[Teacher: What kind of shape is this?]* It's a rectangle.

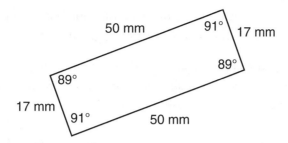

Although the student correctly said that the sides of a rectangle have to be perpendicular, she did not understand that the sides in the above shape are not perpendicular because they meet at angles that are not 90°.

Response 2: Four sides, and 4 right angles. *[Teacher: How about (points to a nonrectangular parallelogram)?]* That looks like a rectangle, but it doesn't have any right angles. I don't think that a rectangle has to have a right angle.

This student seemed to correctly recall the defining property of rectangles (a quadrilateral with 4 right angles). However, because his verbal description was inconsistent with his visual identification of rectangles, he abandoned his correct verbal description and misidentified a nonrectangular parallelogram as a rectangle. Thus, even though the student used the correct language to describe rectangles, his formal reasoning is inadequate because his identification of rectangles was not only inconsistent with his description but also wrong.

Task: A special drawing machine made these shapes *[A–G]*. How are the shapes made by this drawing machine alike? What's the rule?

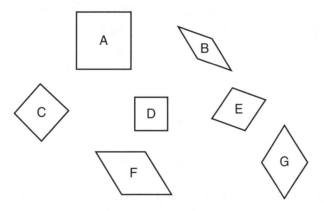

Response: The shape has to be symmetrical, the sides across have to be parallel, and it also has to be equal on all sides, and it also has to have 4 sides. *[Teacher: Which of these shapes can be made by this drawing machine?]*

Levels of Sophistication in Student Reasoning: Geometric Shapes

[Chooses J, K, and L] These two *[K and L]* have 4 equal sides, and are parallel, and are symmetrical. And this one *[J]* has 4 equal sides and is symmetrical.

Although all the properties that the student described for Shapes A–G are formal and correct, he said that Shape K has 4 equal sides, which is incorrect. So his incorrect identification of shapes was inconsistent with his correct formal description.

Task: *A special drawing machine made these shapes. How are these shapes alike? What's the rule?*

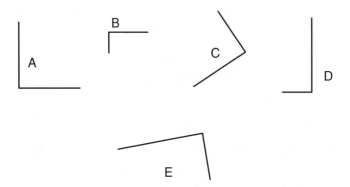

Response: They all look like right angles. *[Teacher: This shape cannot be made by the drawing machine. Why?]*

Because the rule for this one is always straight and straight, and this one's slanted. So this one can't be because it's an acute angle. *[Teacher: And the others were?]* Right. *[Teacher: Which of these shapes can be made by the drawing machine? Why? Why not?]*

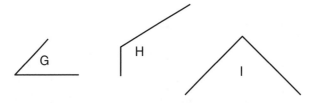

[Circles I] This one *[G]* is an acute angle, it's too small, and this one's *[H]* too big, it's an obtuse angle, but this one *[I]*, if you look at it like this *[turns student sheet so that one ray is parallel to edge of table]* is straight and straight. *[Teacher: And that one is?]* A right angle.

This student correctly uses the formal concepts of right, acute, and obtuse angle to reason about the problems. But his use of informal descriptions like "straight and straight, and this one's slanted" to distinguish right angles from acute angles suggests that his formal ideas are still tightly connected to his informal ideas. This example is included in this section because it illustrates how students in transition from Level 2.2 to Level 2.3 may continue to make sense of formal concepts using informal ideas.

For strategies to help students at Level 2.2, see Chapter 3, page 75.

Level 2.3 Student formally describes shapes completely and correctly.

Students use formal language and concepts to completely and correctly specify classes of shapes, and students' identification of shapes is consistent with their descriptions. Students have made a decided shift away from visually dominated reasoning because the major criterion for identifying a shape is whether it satisfies a set of verbally stated formal properties. So, for example, the term *rectangle* refers to a class of shapes that possesses all the properties the student has come to associate with the set of rectangles (e.g., opposite sides equal and parallel, and 4 right angles).

Students can use and create formal definitions for classes of shapes. However, the shape definitions that students create are not minimal; they list all the properties that they associate with the shape because they do not interrelate properties or recognize that some properties imply other properties.

Progressing to Level 2.3 requires that students understand formal concepts like side-length and angle-measure sufficiently so that they can use the concepts to describe important spatial relationships between shape parts.

EXAMPLES

Task: *Why is this shape not a square?*

Response: Because a square has 4 sides, and this has 5 sides.

Counting geometric parts of shapes such as the sides is not as sophisticated as specifying relationships between parts of shapes (e.g., saying that opposite sides are equal in length or parallel), but it is still property-based reasoning.

Task: *Drawing Machine 1 made these shapes. All shapes that can be made by this drawing machine are alike in some way. How are these shapes alike? What's the rule?*

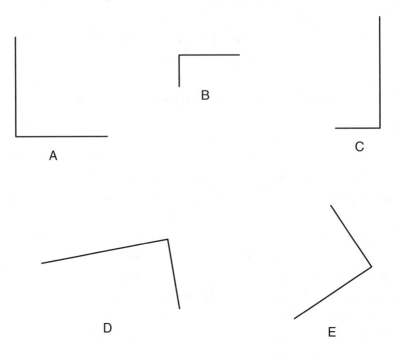

Response: They are all 90° angles. *[Teacher: This shape cannot be made by Drawing Machine 1. Why?]*

Because it is not a 90° angle. *[Teacher: Which of these shapes can be made by Drawing Machine 1? Why? Why not?]*

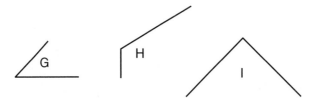

[Circles I] 'Cause if you turn the paper *[rotates student sheet]*, you can see that it *[I]* is a 90° angle.

This student's identification of shapes is completely based on the use of a formal geometric property (having right angles).

Task: *Is this shape a square?*

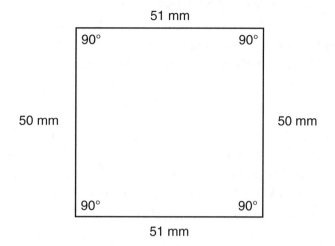

Response: No, I don't think it is a square because a square has all equal sides and this is 50 and this is 51. So I think it would be false.

Task: *Define a rhombus.*

Response: (Borrow, 2000) If you have 4 equal sides and the angles don't always have to be equal, then you could tell it was the rhombus. *[Teacher: How could you tell?]* Because those are the main rules of it. Like, if you had those, you could picture a rhombus in your head.

This student uses two formal geometric properties to describe rhombuses. (The second property, *"the angles don't always have to be equal,"* is usually assumed, not stated, in formal mathematics.)

Task: *Why is this shape not a rectangle?*

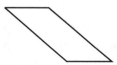

Response: Because it doesn't have right angles.

Task: *Is the shape below a rectangle?*

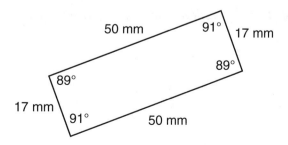

Levels of Sophistication in Student Reasoning: Geometric Shapes

Response: No, because a rectangle has to have 90°; this shape doesn't have all 90° angles, here and here and here and here *[pointing to each vertex]* it has 91, 91, 89 and 89. *[Teacher: Why do you think rectangles have all 90° angles?]* Because they wouldn't be rectangles if they didn't have all 90° angles? *[Teacher: Why wouldn't they be rectangles if they didn't have all 90° angles?]* Because if the angles were too far *[points to the vertices of the figure as he is explaining]*, it would be a parallelogram. If they were too far inward or too far out here, it wouldn't be a rectangle, it would be a parallelogram.

This student not only specifies that rectangles have four 90° angles, he explains why. In saying that if the angles were not 90° they would be "too far inward or too far out," he makes explicit his understanding of the spatial relationship between adjacent sides in a rectangle. He explicitly relates the formal concept of right angle to a special spatial relationship between the sides of the shape.

Task: *Do the clues below guarantee that a figure is a rectangle?*

(1) both pairs of opposite sides parallel

(2) at least one right (90°) angle

Response: They just say one right, 90° angle. It should be four.

This student knows that a rectangle has four 90° angles. But she cannot yet reason that if a quadrilateral has opposite sides parallel and at least one 90° angle, that the quadrilateral must be a rectangle. That reasoning occurs only when students reach Level 3.3. However, her description that a rectangle must have four 90° angles is a correct description of a property of rectangles; it is correct Level 2.3 reasoning about properties of rectangles.

Task: *Tell whether the statement "All squares are rectangles" is true or false.*

Response: False. *[Teacher: What would you say to somebody to prove you are right?]* A square has equal sides *[draws a square]* and a rectangle *[draws rectangle]* doesn't.

This student reasons based on properties. He knows that a square has all sides equal, but he still believes, as do most elementary and middle school students, that rectangles cannot have all sides equal. Although his conclusion is incorrect, his reasoning about properties of squares and properties of rectangles is correct, given that he believes that squares are not rectangles. Understanding that squares are rectangles requires Level 3.4 reasoning.

For strategies to help students at Level 2.3, see Chapter 3, page 96.

LEVEL 3: Student Interrelates Properties and Categories of Shapes[1]

Students explicitly interrelate properties of geometric shapes. For example, a student might say, "If a quadrilateral has opposite sides equal in length, then it has opposite sides parallel." In this case, the student is interrelating Property 1 *"a quadrilateral has opposite sides equal in length"* with Property 2 *"a quadrilateral has opposite sides parallel"* by asserting that when Property 1 is true, Property 2 must be true.

The major factor that varies at Level 3 is the sophistication of students' *justifications* for statements about relationships between properties. These justifications start with empirical associations (when Property X occurs, Property Y occurs), progress to construction-based explanations for why one property "causes" another property to occur, move to logically inferring one property from another, and end with using inference to organize shapes into a hierarchical classification system.

During Level 3, as students' understanding of relationships between properties increases, they organize sets of properties; form minimal definitions for shapes; distinguish between necessary and sufficient sets of properties, and hierarchically organize categories of shapes.

Level 3.1 Student uses empirical evidence to interrelate properties and categories of shapes.

Students use empirical evidence (observations of examples) to conclude that if a shape has one property, it has another property. They make this conclusion because in all the examples they have seen or can imagine, when one property occurs, so does the other property.

EXAMPLES

Task: *True or False? If a quadrilateral has opposite sides equal and at least one right angle, then the quadrilateral is a rectangle. Prove your answer or tell why your answer is correct.*

Response: True. All the shapes that I can think of with opposite sides equal, when they have a right angle, they have all right angles.

Task: *True or False? If a quadrilateral has 4 right angles, then the opposite sides of the quadrilateral must be parallel. Prove your answer or tell why your answer is correct.*

Response: I think it's true. If they are all right angles, then I think the opposite sides are parallel. *[Teacher: Tell me why.]* Because if all the angles are right angles, the sides have to be parallel . . . . That's like what always happens . . . . I haven't really found any shape where it wouldn't be true.

---

[1] Level 3 sublevels are based on collaborative work between Battista and Borrow. Some of the examples in this section are based on those reported by Borrow (2000).

Task: *Which is the best way to define a kite, by saying it is a quadrilateral with "at least 1 line of symmetry angle to angle" or a quadrilateral with "at least 2 sets of adjacent sides congruent"?*

Response: You could do either. *[Teacher: Why?]* We worked on the computer a while and couldn't find any shapes *[made by a computer manipulable Kite Maker]* that proved it wrong. If you have those 2 adjacent side lengths equal, it automatically makes a line of symmetry. And if you have at least 1 line of symmetry, the adjacent line lengths have to be equal. If you have one, it automatically causes the other to happen.

These three students justified their property inferences based on empirical observations—for all the examples the students could think of, when one property occurred, the other property occurred.

For strategies to help students at Level 3.1, see Chapter 3, page 98.

Level 3.2 Student analyzes shape construction to interrelate properties and categories of shapes.

Students analyze how a shape, like a rectangle, can be constructed, one part at a time, to conclude that when one property occurs, another property must occur. Students make drawings, build models, or use imagery to analyze the part-by-part construction. For instance, a student might conclude that if a quadrilateral has 4 right angles, its opposite sides are equal because if you draw a rectangle by making a sequence of perpendiculars, the sides *must* be equal. The student explicitly sees visually that making a pair of opposite sides unequal causes some of the angles to *not* be right angles.

EXAMPLES

Task: *True or False? If a quadrilateral has opposite sides equal and at least one right angle, then the quadrilateral is a rectangle. Prove your answer or tell why your answer is correct.*

Response 1: It's true. Because, say, there's a 90° angle right here *[draws a 90° angle; see Figure 1]*. This side right here *[left side]* has to be equal to this side right there *[draws a vertical line segment opposite the left side that is already drawn; see Figure 2]*. And it has to be lined up with the top of the *[given]* right angle, like across here *[extends the partial segment across the top to meet the right side; see Figure 3]*. If it's not lined up, it won't be a right angle here *[pointing to the angle in the upper right of the figure]*. And you get right angles on the bottom because this side across the bottom has to be equal to the top side *[draws the bottom side; see Figure 4]*. So it fits in to make right angles; if it's not the same as the top, then you don't get right angles. So you get a rectangle, which has four 90° angles.

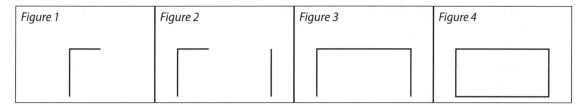

Based on her analysis of the *process* of constructing a quadrilateral with the given properties, part-by-part, this student argued that if you construct a quadrilateral having one right angle and opposite sides equal you *must* get a rectangle, which has four 90° angles. (Note that there are logical gaps in her argument.)

Response 2: I'm drawing a right angle *[draws a right angle]*.

Make these two lines equal *[draws a third segment and puts dashes on the two segments that are supposed to be equal in length]*. Connect the second and third sides with an angle that's not a right angle.

If you connect those two at the bottom *[draws the fourth side]*, then these opposite sides *[pointing to the top and bottom sides]* won't be equal.

So, if you want to make opposite sides equal and at least one right angle, you have to make all the angles right angles. *[She draws a right angle, then a third segment perpendicular to one of the extended sides and congruent to its opposite, then a fourth segment closing the figure.]*

This student's part-by-part construction convinced her and enabled her to argue intuitively (using an informal but incomplete proof by contradiction), that if you construct a quadrilateral with the properties "opposite sides equal" and "at least one right angle," then the other angles must be right angles.

For strategies to help students at Level 3.2, see Chapter 3, page 102.

Level 3.3 Student uses logical inference to relate properties and understand minimal definitions.

Students make logical inferences about geometric properties, starting to develop the ability to relate properties as mathematicians do. For instance, a student whose definition of a rectangle is "4 right angles and opposite sides equal" understands the logic in the statement, "a square is a rectangle because (a) it has 4 right angles, like rectangles; and (b) since it has 4 equal sides, it has opposite sides equal, like rectangles."

Students form and accept minimal definitions for shapes—that is, correct definitions that list a *sufficient* set of properties—rather than all the properties that they know. For instance, a student reasoning at Level 3.4 might know that rectangles have 4 right angles, opposite sides equal in length, and opposite sides parallel, but might define a rectangle as "a quadrilateral having opposite sides equal and 4 right angles." The student accepts this definition because she knows she can logically deduce that if a quadrilateral has 4 right angles, it must have opposite sides parallel.

Students' reasoning is "locally logical" in that they string together logical deductions based on "assumed-true" propositions—that is, propositions they accept as true based on their experience or intuition, or because their teacher or textbook says they are true. Thus, students at this level use logic, but they do not question the starting points for their logical analyses.

Although students who are reasoning at Level 3.3 possess the ability to make the inferences needed for hierarchical classification of shapes (e.g., all squares are rectangles), these students do not use their deductive ability to logically *reorganize* their conceptual network about shapes. They have difficulty accepting a logical hierarchical shape classification system (e.g., students still resist the notion that a square is a rectangle even though they can see the logic in such a statement).

EXAMPLES

Task: *How are these shapes alike?*

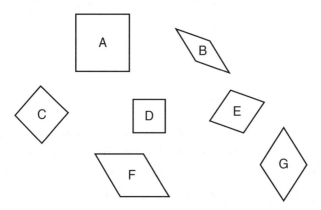

Response: They all have 2 lines of symmetry. The squares go up like this and this *[draws a horizontal and vertical line of symmetry on Shape A]* and these *[draws diagonals on F]* go like this and like this. But the squares can have more than 2 lines of symmetry, so they *[Shapes A–G]* should have *at least* 2 lines of symmetry because this Shape *[F]* can only have 2 lines of symmetry.

This student made one of the simplest kinds of logical connections of properties; he connected the properties "have 2 lines of symmetry," and "have more than 2 lines of symmetry" with the statement "have at least 2 lines of symmetry."

Task: John claims that if a figure has 4 equal sides, it is a square. Which figure might be used to prove John is wrong?

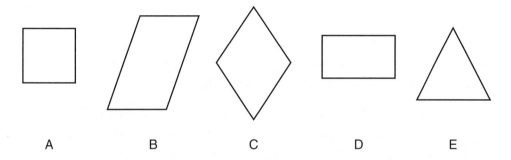

A   B   C   D   E

Response: *[Circles C]* This shape could be used to show John is wrong because it is a rhombus, so all its sides are equal. *[Teacher: Why does that prove that John is wrong?]* He says if all sides are equal then it has to be a square. But the 4 sides are equal on this *[C]* and it is not a square.

This student realizes that if he finds a shape that has all side lengths equal but is not a square, he has found a counterexample that proves John's statement is incorrect.

Task: (Borrow, 2000). *Evaluate the following definition: "A square is a quadrilateral with 4 congruent sides and at least 1 right angle." Does this definition give enough information to guarantee that a shape is a square?*

Response: Yes. The only other shape that we know of that has 4 equal sides is the rhombus. And if you know there is one right angle in a rhombus, then *[pause]*. Well if you think of a rhombus, it has the opposite angles equal. So if one angle is 90°, then you say there are 2 angles that are 90°, and that equals 180°. But the other 2 angles must add up to 180° because the total is 360°. So, because these 2 angles are equal, they have to each be 90°. So it must be a square.

Based on ideas that she accepted as true (rhombuses have opposite angles equal, the sum of the angles in a quadrilateral is 360°), the student logically deduced that if a quadrilateral has 4 congruent sides and at least 1 right angle, then it must have four 90° angles, and so must be a square.

Task: *Give a definition for a rectangle that does not list all of its properties.*

Response: I know that rectangles are quadrilaterals with 4 right angles, opposite sides equal, and opposite sides parallel. But my definition is: A rectangle is a quadrilateral that has 4 right angles and opposite sides equal. *[Teacher: Why don't you include "opposite sides parallel"?]* Because I know that if a quadrilateral has 4 right angles, it has opposite sides parallel. Remember we learned in class that 2 lines are parallel if their inside angles make 180°. So here's my rectangle and it has 4 right angles *[draws rectangle ABCD]*.

Because angles A and D are 90°, they add to 180°, so the top and bottom sides are parallel. And because angles A and B are each 90°, the other 2 sides are parallel.

This student uses his ability to logically deduce one property of rectangles from other properties to construct a definition of rectangles that does not list all the properties of rectangles that he knows. Furthermore, he is able to justify why his "smaller" but sufficient definition implies all the properties he thinks rectangles should have.

Task: *Students are trying to decide on the truth of the statements (a) squares are rectangles, and (b) rectangles are squares.*

Response: You have mentioned that opposite sides are parallel and equal in a rectangle. It's the same way with a square except that all sides are equal. But all the sides being equal means that opposite sides are equal. So a square in the sense that you're saying is still a rectangle, but a rectangle is not a square. . . . Well, one rectangle is a square, the rectangle with equal sides.

This student sees that, for quadrilaterals, *"all the sides being equal"* implies that *"opposite sides are equal."* Based on this inference, she comprehends the logic inherent in the mathematical classification that squares are rectangles. But her comment "in the sense that you're saying" suggests that she has not yet embraced this organization as her own.

Task: *How are squares and rectangles related?*

Response: Why don't you call a rectangle a square with unequal sides, instead of saying that a square is a special kind of rectangle? *[Teacher: Remember, we defined a rectangle as a shape that has 4 right angles and opposite sides parallel.]* If you use your definition, then a square is a rectangle because it does have right angles and opposite sides parallel.

This comment, like those of the previous student, clearly indicates an ability to follow the logic in the mathematical classification that squares are rectangles. But neither student has yet adopted the logical-mathematical classification of quadrilaterals—each still clings to the idea of each type of quadrilateral as separate. Students feel comfortable classifying squares as rectangles only when they personally accept the usual hierarchical shape classification system. For these students to move to Level 3.4, they must personally reorganize their definitions of shapes into a hierarchical classification scheme. This hierarchical reorganization does not automatically result from the ability to follow and make logical deductions. Furthermore, if we want students to make personal sense of and accept this organization, we cannot force it on them.

For strategies to help students at Level 3.3, see Chapter 3, page 106.

Level 3.4 Student understands and adopts hierarchical classifications of shape classes.

Students use logical inference to *reorganize* their classification of shapes into a logical hierarchy. For example, students see and *accept* the notion that all squares are rectangles, all rectangles are parallelograms, and all parallelograms are quadrilaterals. It becomes not only logically clear why a square is a rectangle, but because students feel that hierarchical classification is a necessary part of sound reasoning, they fundamentally restructure their classification of shapes. Furthermore, they give logical arguments to justify their hierarchical classifications.

Students can use definitions to logically organize shapes by their properties. For instance, students recognize that a rectangle can be defined as a parallelogram with at least 1 right angle, or a square can be defined as a rhombus with at least 1 right angle, or a rhombus can be defined as a parallelogram with all sides equal. These definitions enable students to reorganize their conception of the set of quadrilaterals into the hierarchical classification shown on the following page.

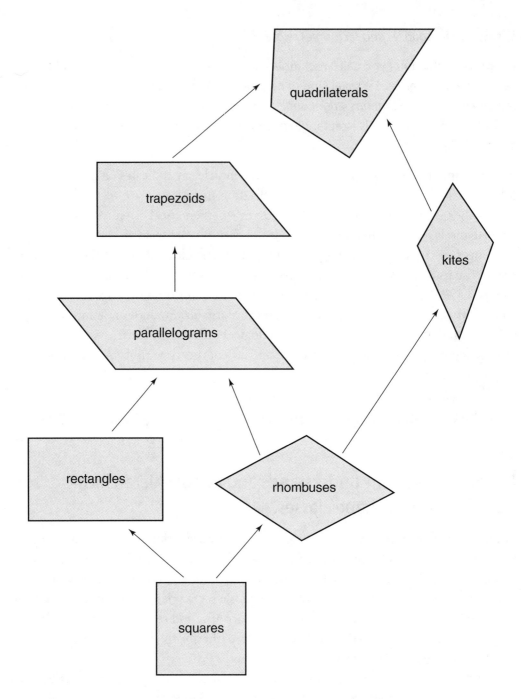

Students' use of logic to draw conclusions provides them with a new way to accumulate knowledge. New knowledge can now be generated not merely through empirical or intuitive means but through logical deduction.

Students also start to appreciate how showing a property true for one class of shapes means that the property is true for all the subclasses of the shape (this is the power of hierarchical classification). For instance, we might discover that the diagonals of a parallelogram bisect each other. From that discovery, we can deduce that the diagonals of rectangles, squares, and rhombuses must also bisect each other.

EXAMPLES

Student 1: I think that when you say that a shape has the properties for a rectangle, it could always be a square. Like the traditional properties that you think of, or at least what I think of for a rectangle, have always been that the opposite sides are equal and it has four 90° angles. But that could be a square because a square has opposite sides equal and four 90° angles. So a square is a *kind of* rectangle.

Student 2: A square and a rhombus are always kites because they always have all sides equal, which means they have 2 pairs of adjacent sides equal that you need to be a kite.

Student 3: A rhombus always has 2 sets of parallel sides, so it is always a parallelogram. Also, a rhombus is always a kite because all its sides are equal, which means adjacent sides are equal.

Student 4: All squares are kites because all squares have 2 lines of symmetry from angle-to-angle, and a kite has to have at least one line of symmetry angle-to-angle.

All 4 of these students accept parts of a hierarchical classification of shapes and use logical deduction to justify their hierarchical classifications.

For strategies to help students at Level 3.4, see Chapter 3, page 108.

LEVEL 4: Student Understands and Creates Formal Deductive Proofs

Students understand and can construct formal geometric proofs, as is required in traditional high school geometry courses. That is, within an axiomatic system, they can meaningfully produce a sequence of statements that logically justifies a conclusion as a consequence of the "givens." They recognize the difference among undefined terms, definitions, axioms, and theorems. At Level 4, students understand that to genuinely validate their reasoning, they must provide formal proofs.

EXAMPLES

Task: *Prove: If a quadrilateral has 4 right angles, then the opposite sides of the quadrilateral must be parallel.*

Response: We are given that all 4 angles of quadrilateral ABCD are right angles; that is, each measures 90°.

Because the sum of the measures of angles A and B equals 180°, side BC is parallel to side AD. This is because when lines BC and AD are cut by the transversal AB, the same side interior angles, angles A and B, are supplementary, which by one of our theorems, implies that BC is a parallel to AD.

Using similar reasoning, because the sum of the measures of angles B and C equals 180°, side AB is parallel to side DC. So opposite sides of the quadrilateral are parallel.

Note that this proof is similar to the argument provided in the last example for Level 3.4. The only difference is that this proof references a number of formal concepts from an axiomatic system: transversals, same-side interior angles, supplementary, and a specific theorem.

Task: *Prove: If a quadrilateral has opposite sides equal and at least one right angle, then the quadrilateral is a rectangle.*

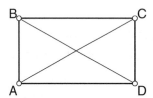

Response: We are given that opposite sides of quadrilateral ABCD are congruent (equal in length). This means that AB is congruent to DC and BC is congruent to AD.

We know that BD is congruent to BD (by the reflexive property of congruence).

So, by the Side-Side-Side congruence theorem, triangle BAD is congruent to triangle DCB.

We are also given that one angle is a right angle; suppose this angle is angle BAD. Because of our definition that corresponding parts of congruent triangles are congruent, angle DCB must also be a right angle. So we now know that angles BAD and DCB are both right angles, and thus sum to 180°.

Using a similar argument, we get that triangle ABC is congruent to triangle CDA, and that angle ABC is congruent to angle CDA.

Because of our theorem that the sum of the angles in a quadrilateral is 360°, and the measures of angles BAD and DCB sum to 180°, we know that angles ABC and CDA sum to 180°, and are thus both right angles. So all the angles of quadrilateral ABCD are right angles.

Because angle ABC plus angle BAD equals 180°, side BC is parallel to side AD. This is because of our theorem that states that when lines BC and AD are cut by the transversal AB, and the same side interior angles, angle ABC and angle BAD, are supplementary, then BC is a parallel to AD.

Using similar reasoning, because angle ABC plus angle BCD equals 180°, side AB is parallel to side CD.

So the opposite sides of quadrilateral ABCD are parallel.

Task: *Prove that the diagonals of a rectangle are congruent.*
   Given: ABCD is a rectangle.
   Prove: Diagonal AC is congruent to diagonal DB.

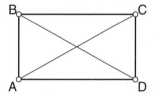

Response:

AB is congruent to DC because opposite sides of a rectangle are congruent.

AD is congruent to AD because of the reflexive property of congruence.

Angle BAD is congruent to angle CDA because all angles in a rectangle are right angles and so are congruent.

Triangle BAD is congruent to triangle CDA because of the Side-Angle-Side congruence theorem.

Therefore side AC is congruent to side DB because corresponding parts of congruent triangles are congruent.

Using CBA Levels to Develop a Profile of a Student's Reasoning About Geometric Shapes

CBA assessment tasks are designed to help you assess levels of reasoning, not levels of students. Indeed, a student might use different levels of reasoning on different tasks. For instance, a student might operate at a higher level when using physical materials than when she does not have physical materials to support her thinking. Also, a student might operate at different levels on shapes that are familiar to her as opposed to shapes that are totally new to her. Rather than attempting to assign a single level to a student, you should analyze a student's reasoning on several assessment tasks, then develop an overall profile of how she is reasoning about the topic.

To develop a CBA profile of a student's reasoning, note which CBA assessment tasks you give to the student, the date, and what CBA level of reasoning the student used on each task. Note whether the student used physical materials (P), drew pictures (D), used paper-and-pencil to write computations (PP), did the problem strictly mentally (M), or used the computer (C). Record whether the student answered the questions correctly or not (C or I). Some teachers also note whether a student was being guided by a teacher (T) or worked on the task without any help (WH). These annotations can be quite important in monitoring student progress. For instance, if in the initial assessment a student uses one level of reasoning with the help of physical materials but in a subsequent assessment the student uses the same level of reasoning implemented mentally, the student has made considerable progress.

As shown below, a CBA profile provides an excellent picture of student reasoning that can be monitored throughout the school year.

CBA Reasoning Profile

RH, Grade 5, Geometric Shapes

Step 1: *Record What RH Did on CBA Problems*

1. A person has a special drawing machine that makes these shapes. All the shapes made by this drawing machine are alike in some way. Your job is to figure out what kind of shape this drawing machine makes.

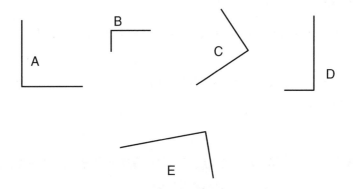

How are these shapes alike? What's the rule?

RH: *The shapes are alike because they all have right angles. They are all straight lines. They are not curved. And I think you could make a square or a rectangle with them.*

Teacher: *What did you mean by a right angle?*

RH: *The angle is like straight. Like this is a right angle* [draws a right angle] *and this would not be a right angle* [draws an acute angle] *because the lines are not. A right angle is like a 90° angle and that* [pointing to the second angle she drew] *would not be a 90° angle because the lines would be too close together or they would be too far apart.*

This shape cannot be made by the drawing machine. Why?

    RH: *Because it is not a right angle.*

    Teacher: *What would the rule be for the drawing machine to make these shapes?*

    RH: *They have to be right angles.*

Which of these shapes can be made by the drawing machine? Why? Why not?

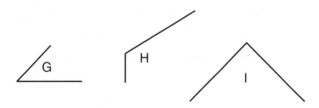

    RH: *I think this one* [circling Shape I]. *Because, again, it is a right angle. And these ones* [G, H], *no, because the lines are either too far out or too far in.*

Because RH describes the shape made by the drawing machine as right angles, but also refers to the sides of angles as being "straight" and "too far out or too far in," her reasoning is Level 2.2.

  2. All the shapes made by another drawing machine are alike in some way. Can you figure out what kind of shapes this drawing machine makes?

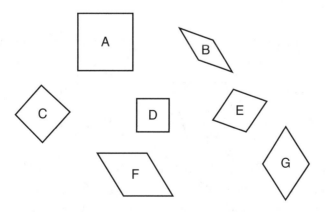

Levels of Sophistication in Student Reasoning: Geometric Shapes

How are these shapes alike? What's the rule?

RH: *I think they are the same shape almost. But if you would stretch that angle down to match this one* [drawing an arrow from the bottom left angle of Shape B] *and stretch this to go out and stretch this one. That one would actually go in. And then it would make a square.*

RH: *I am thinking it is a right angle, but I think I am wrong.*

Teacher: *Why do you think you are wrong?*

RH: *The right angle is tilted. Like this right angle on this square* [pointing to the lower left right angle on Shape A], *if you put it down here* [pointing to the lower left angle of Shape F], *it wouldn't match up. But I don't think it is a right angle because that one is crooked. Not crooked, too far out* [pointing to the lower, left angle on Shape F]. *Oh, I know how they are alike. The opposite sides are parallel.*

This shape cannot be made by this drawing machine. Why?

RH: *Oh, not all the sides are the same length. Two sides are bigger.*

Teacher: *Now what do you think the rule is for this drawing machine?*

RH: *Each side has to be the same length.*

Teacher: *Anything else?*

RH: *No.*

Which of these shapes can be made by this drawing machine? Why or why not?

RH: *The last one* [Shape L] *because it is the same as this one* [Shape D]. *And I think it could also be this one* [Shape J]. *And that is it. Because their sides are the same length. And these* [I and K] *are not the same length. One side is longer than the other.*

RH's initial response was a mixture of Levels 1.2 and 2.1 reasoning. But by the end, her response was at Level 2.3. So, overall, we might say that her reasoning was Level 2.2 because she used a mixture of formal and informal reasoning, mostly about parts of shapes.

3. Here is a new drawing machine. All shapes made by this drawing machine are alike in some way. Can you figure out what kind of shapes this drawing machine makes?

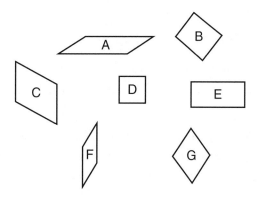

How are these shapes alike? What's the rule?

RH: *Opposite sides are parallel. They are all quadrilaterals. All have 4 sides.*

This shape cannot be made by the drawing machine. Why?

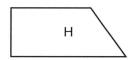

RH: *Because the opposite sides are not parallel.*

Teacher: *Which ones?*

RH: *These two are not parallel* [drawing a mark on the nonparallel lateral sides] *and these two are* [pointing at the top and bottom sides].

Teacher: *What do you think the rule is?*

RH: *They have to be parallel, opposite sides.*

Which of these shapes can be made by the drawing machine? Why or why not?

  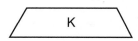 

Levels of Sophistication in Student Reasoning: Geometric Shapes

RH: *That one [Shape I] because it is the same as this shape [pointing to Shape E] but it is smaller and it is tilted a different way, slanted. And this one [circling Shape J] because it is like this one [pointing to Shape G] but it is smaller. And yeah, it is smaller. And that is it.*

Teacher: *Why did you not circle this one [pointing to Shape K]?*

RH: *One of the pairs of opposite sides is not parallel.*

Teacher: *And how about the fourth shape [L]?*

RH: *No, because sides are both slanted in to each other. And none of the others are [pointing to Shapes A–G and Shapes I and J].*

We can classify RH's reasoning on this task as almost Level 2.3 reasoning. However, because of her last statement "because sides are both slanted in to each other," she is still at Level 2.2.

4. Here is a new drawing machine. It also makes shapes according to a rule. So all shapes made by this drawing machine are alike in some way. Can you figure out what kind of shapes this drawing machine makes?

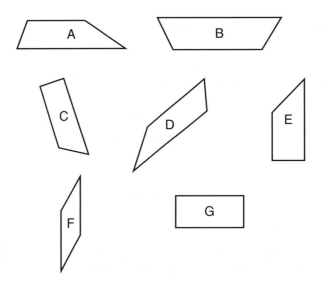

How are these shapes alike? What's the rule?

RH: *They are all quadrilaterals. Not all of them, all of them but 2 are trapezoids.*

Teacher: *Which two?*

RH: *F and G are not trapezoids.*

Teacher: *You said a lot of them are trapezoids. What makes them trapezoids?*

RH: *Two of the sides are not parallel; 2 of the opposite sides. And the 2 angles go out.*

This shape cannot be made by the drawing machine. Why?

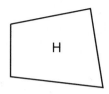

RH: *The sides are slanted in. Two of the sides are slanted in, these two* [pointing to the top and bottom of Shape H].

Which of these shapes can be made by the drawing machine? Why or why not?

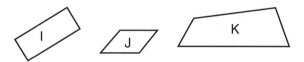

RH: *These two* [Shapes I and J] *because this one* [Shape I] *is almost like this rectangle* [pointing to Shape G]. *And then this shape* [Shape J] *is almost like that shape* [Shape F] *only wider. And I don't think that shape* [Shape K] *is because; that might be able to, I think.*

Teacher: *Why?*

RH: *It is almost a trapezoid.*

Teacher: *Almost. What makes it not a trapezoid?*

RH: *One of the sides is smaller.*

Teacher: *Let me see.*

RH: *This line* [referring to the top of Shape K], *if you draw a straight line* [drawing a segment from the top right vertex of Shape K and making it parallel to the bottom side] *it doesn't go completely to the other angle.*

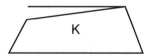

Teacher: *If this was a straight line* [referring to the segment she drew on top of Shape K], *and this angle did come up to here?*

RH: *It would be a trapezoid.*

Teacher: *And what does this trapezoid have that makes it like other trapezoids?*

RH: *Like one pair of the opposite sides are parallel and then maybe the angles can be the same.*

Teacher: *And what did you say about these shapes* [pointing to Shapes A–G]?

RH: *They are all trapezoids.*

Teacher: *Oh.*

RH: *Not all.*

Teacher: *Do you want to change the rule at all now?*

RH: *I'll stick with the sides don't curve.*

RH has not yet clearly formulated the definition of trapezoids in terms of properties. Her statements, "like one pair of the opposite sides are parallel and then maybe the angles can be the same" and "the sides are slanted in" indicate that she is close to a correct definition. But it is insufficient to properly specify trapezoids. Also, as is typical of students who are reasoning at Level 2, she has difficulty recognizing that Shapes F and G (a parallelogram and a rectangle) are trapezoids (in her school, they used the inclusive definition, so parallelograms are trapezoids).

5. Circle each triangle.

RH circled Shapes 1, 4, 5, and 6.

Teacher: *Why did you say that Shape 1 was a triangle?*

RH: *Because it has got 3 sides, same as* [Shapes] *4, 5, and 6. Each side is straight and it has 3 points.*

Teacher: *Why did you say Shape 2 was not?*

RH: *Because it has 4 points and also 4 sides.*

Teacher: *And why did you say Shape 3 was not?*

RH: *Because it is kind of round, like it is not perfectly straight lines. Like it goes in on some places and out on the others.*

Teacher: *And why did you say Shape 8 was not?*

RH: *I am not sure. Because the bottom is not straight, like all the other bottoms on the other triangles are. It is slanted. It is kind of like a triangle sideways. This would be the bottom if you stood it up* [pointing to the shortest side of Shape 8]. *The lines, they don't go like that. They go the same way. They don't go in the opposite directions on each side.*

Teacher: *Show me where they do that.*

RH: *That could be a triangle. I think it is a triangle.*

Teacher: *Why do you think it is a triangle?*

RH: *It looks like a triangle because it does have 3 points. And it might be a triangle I have never heard of. Like, I recognize these* [pointing to the triangles she circled]. *It is just that I have seen them previously.*

Teacher: *And why did you say* [Shape] *9 was not?*

RH: *It doesn't have a bottom.*

Teacher: *And why did you say Shape 7 was not?*

RH: *Because it goes in on the bottom and that would make it have 4 points.*

Because RH has difficulty classifying shapes strictly based on properties *[Shape 8]*, she is at Level 2.2.

6. Circle each rectangle.

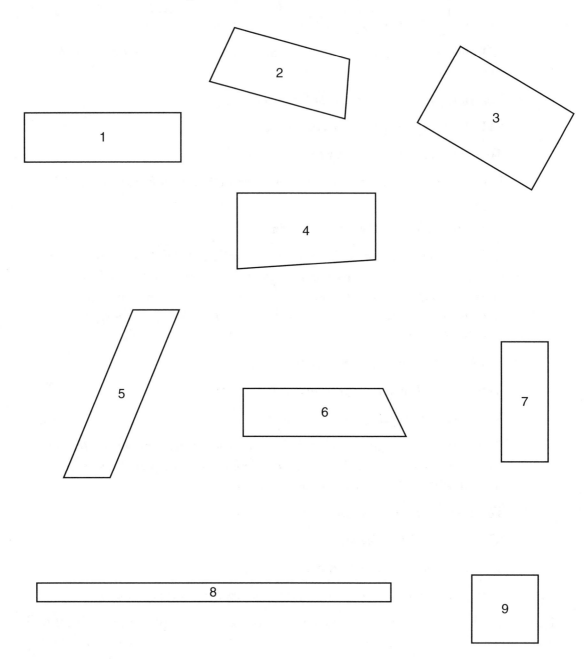

RH circled figures 1, 3, 7, and 8.

Teacher: *Why did you say that Shape 1 was a rectangle?*

RH: *It has got 4 points. The opposite sides are parallel. And the opposite sides are the same length. And that is it.*

Teacher: *And why did you say Shape 4 was not?*

RH: *Like on a rectangle, all the lines have to be like straight across. But it [Shape 4] goes up. So that would make not all sides the same length.*

Teacher: *And why did you say that Shape 6 was not?*

RH: *Because this side is not straight* [circling the right side of Shape 6]. *So it would not be parallel with the opposite of it, with this one* [pointing to the left side].

Teacher: *And why did you say Shape 9 was not?*

RH: *Because all the sides are the same length. On rectangles not all sides can be the same length.*

Teacher: *Describe exactly how you decide if a shape is a rectangle or not.*

RH: *It has 4 sides, which is a quadrilateral. The opposite sides are parallel. The sides are not the same length. That's it.*

Even though RH correctly identifies all rectangles (except the square, which comes only at Level 3), she did not know all the properties of rectangles. She still does not specify right angles.

7. Circle T for True, or F for False, for each statement about the shape below. (Circle "Can't tell" if you think there is no way to know for sure.)

   (a) The shape is a square.   T   F   Can't tell
   (b) The shape is a rectangle.   T   F   Can't tell
   (c) The shape is a parallelogram. T   F   Can't tell
   (d) The shape is a rhombus.   T   F   Can't tell

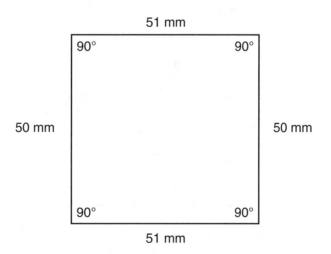

RH said a is true and b, c, and d are false.

RH: [Problem] *(a) the shape is a square. It is true. It is not a rectangle, so that would be false. It is not a parallelogram, and it is not a rhombus.*

Describe how you would convince someone that your answers are correct.

RH: *Because in a rectangle, the sides are not the same length, like a square. In a parallelogram the sides are not straight going up and down, or they are not straight side to side. And the same in a rhombus, the sides aren't straight up and down or straight sideways.*

Teacher: *This* [the given figure] *is not a parallelogram because the sides are?*

RH: *These are straight and these are straight* [pointing to two adjacent sides on the given figure].
Yeah. [pause] *And in a rhombus and parallelogram it doesn't have 90° angles.*

Teacher: *Can any parallelograms have 90° angles?*

RH: *I don't think so.*

RH's reasoning clearly shows that she is not correctly using formal properties to classify shapes, consistent with her previous Level 2.2 reasoning. In classifying the given shape, she ignores the fact that it does not have all sides equal. And she describes characteristics of parallelograms and rhombuses informally as "the sides are not straight going up and down." She does mention 90° angles in reference to straightness, so she is, characteristic of Level 2.2, starting to reformulate some of her informal properties in terms of formal concepts. (Indicates the need for more work on angles.)

8. Circle T for True, or F for False, for each statement about the shape below. (Circle "Can't tell" if you think there is no way to know for sure.)

    (a) The shape is a square.   T   F   Can't tell

    (b) The shape is a rectangle.   T   F   Can't tell

    (c) The shape is a parallelogram.   T   F   Can't tell

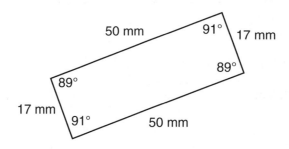

RH circled F for Problem (a), T for Problem (b) and F for Problem (c).

RH: *I know it is not a square because the sides are not the same length. The shape is a rectangle because it has four 90° angles and the sides are not the same length. Wait, it doesn't have 90° angles. Skip the 90° angles.*

Teacher: *What is that?*

RH: *I don't think the 90° angles. I can't tell for rectangle.*

Teacher: *Why can't you tell?*

RH: *Because for rectangle I thought 90° angles on all sides. And it is not the same length. The opposite ones.*

Teacher: *And what is the case here?*

RH: *They are not 90°.*

Teacher: *So what does that mean to you if in the picture they are not?*

RH: *It would be false.*

Teacher: *Why are you thinking it would be false?*

RH: *Because it doesn't have a 90° angle. And now I know that, so I can tell that it is false.*

Teacher: *And how about for a parallelogram?*

RH: *I think it is false.*

Teacher: *Why do you think it is false?*

RH: *In a parallelogram the sides are going the same way. They are going slanted.*

Teacher: *And what about in this* [the given] *shape?*

RH: *They don't go one way. In a parallelogram the 2 sides go one way. On this* [given figure] *they kind of like don't go one way. I think they are straight.*

RH's reasoning about rectangles is starting to move toward Level 2.3, but remains at Level 2.2 about parallelograms.

9. Tell whether the statement in the box below is true or false.

> All squares are rectangles.

RH: *I had this in my class, and I forgot. I forget which way it is. I think it is true. Because the square has a 90° angle, same as a rectangle. That is it. I am kind of tricked on this one.*

Teacher: *You are thinking squares have 90° angles and so do rectangles.*

RH: *It could be the other way around too. That rectangles are squares.*

As is typical of students in Level 2, RH has no substantive understanding of hierarchical classification. The best she can do is resort to rote memorization.

10. I'm thinking of a closed figure with 4 straight sides.

    It has 2 long sides and 2 short sides.

    The 2 long sides are the same length.

    The 2 short sides are the same length.

    What shape could I be thinking of?

Could it be a triangle?	Yes	**No**
Could it be a square?	Yes	**No**
Could it be a rectangle?	**Yes**	No
Could it be a parallelogram?	Yes	**No**

Levels of Sophistication in Student Reasoning: Geometric Shapes

RH: *A rectangle. I don't think it could be a triangle since a triangle only has 3 and you said it has 2 long and 2 short, which equals 4. Could it be a square? No, because in a square all the sides are like equal length. So it really couldn't be a square. And in a rectangle it matches the description. It has 2 long and 2 short sides and the 2 long sides are the same length and the 2 short sides are the same length. It could not be a parallelogram because a parallelogram* [pause]. *I just don't think a parallelogram can match the description because the sides are slanted either one way or the other.*

Again, RH has difficulty thinking of shapes strictly in terms of properties, so she is still reasoning at Level 2.2. For instance, the described shape could be a kite—the 2 long sides could be adjacent instead of opposite.

Step 2: *Construct a CBA Levels Summary Chart for RH*

Task	Level	Correct/Incorrect	Comment
1	2.2	C	
2	2.2	C	Mixture of 1.2, 2.1, 2.3
3	2.2	C	
4	2.2	I	
5	2.2	I then C	
6	2.2	Incorrect (square) C*	RH's response is correct for Level 2, but incorrect for Level 3 where she would understand that squares are rectangles.
7	2.2	I	
8	2.2	I	I then C for rectangle, still incorrect for parallelogram.
9	2	C then I	Resorts to rote memorization.
10	2.2	I	

Summary Diagnosis: RH is wavering between informal and formal reasoning about shapes' properties. In many instances, she starts her reasoning with informal statements, but on further reflection or teacher questioning, sometimes she can reformulate her reasoning in terms of formal concepts. Also, RH has not reached the point where she classifies shapes strictly in terms of their formal properties. Note, however, that when RH has a chance to further reflect on the ideas, she already is reformulating some of her informal reasoning in terms of formal concepts. So instruction should provide plenty of opportunities for RH to extend her reflection on shapes.

Step 3: Develop a Set of Goals and Recommendations for Instruction

Goal. The overall goal is to help RH reformulate her shape definitions in terms of formal concepts such as angle measure, side length, and parallelism.

Subgoal. To help RH reach Goal 1, instruction should first focus on helping RH deepen her understanding of formal concepts so she is better able to use them to conceive of informal ideas.

RECOMMENDATIONS

Recommendation 1. Sharpen and increase RH's use of angle measure to understand the angular relationship possibilities when two line segments intersect.

Task a: *Examine how the intersections of line segments on the left (pictures of nonperpendicular segments) differ from the intersections of line segments on the right (pictures of perpendicular segments). How are they different? What is special about how the segments on the right intersect? Can you state one thing that clearly distinguishes the 2 sets of line segments? How can you use angle measure to clearly distinguish the 2 sets of line segments? If a student said that the segments in the intersections on the left were not straight, and those on the right were straight, how could you explain this idea using angles measurements? (Because RH is at Level 2, do not deal explicitly with hierarchical classification at this point in time.)*

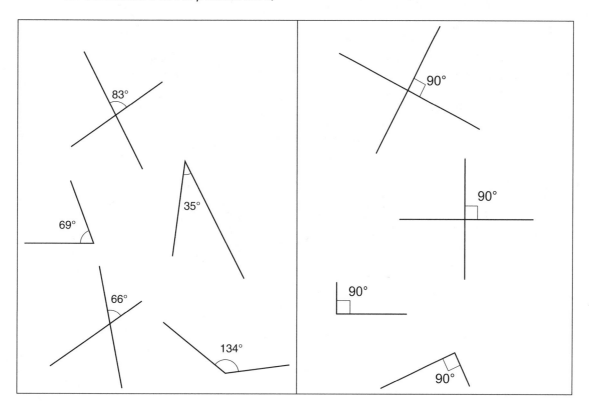

Task b: *Use what you discovered in Task a to describe how the intersections of sides in rectangles and (nonrectangular) parallelograms differ (give students a sheet with both types of figures drawn, with angle measurements showing—see examples below). In rectangles, all the angles are right angles. In nonrectangular parallelograms, the angles are not right angles or 90°.*

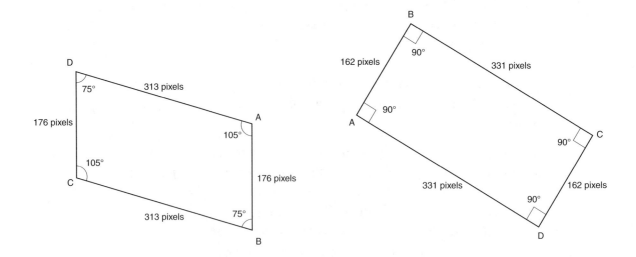

Task c: *Have students use measured Rectangle and Parallelogram Makers in dynamic geometry to try to make various shapes, and explain why shapes can or cannot be made (encourage descriptions that include specification of angle measurements). See Chapter 3.*

Measured Rectangle Maker

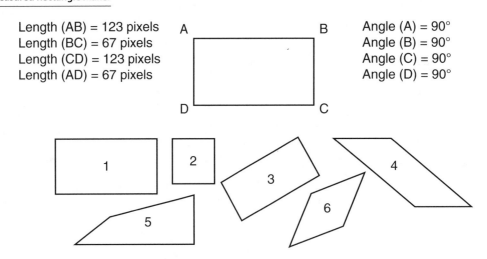

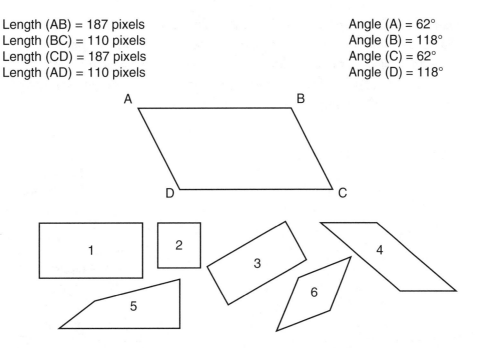

RH should do similar activities with measured Parallelogram, Rhombus, and Kite Makers.

Recommendation 2. Have RH discuss what is meant by parallel. Often, students believe that two line segments are parallel only if they are parallel and equal.

Task a: *Which pairs of segments are parallel?*

Measured Parallelogram Maker

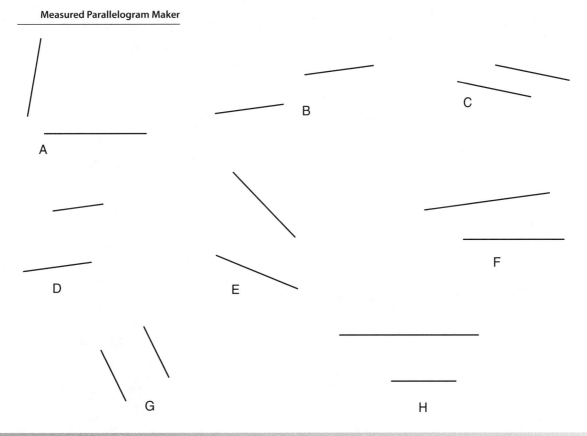

Levels of Sophistication in Student Reasoning: Geometric Shapes

Task b: *Have RH examine parallelograms and nonparallelogram trapezoids and discuss how their sides and angles compare. Which shapes are parallelograms, which are trapezoids but not parallelograms? How do the angle measurements of parallelograms and trapezoids compare? How do the side lengths of parallelograms and trapezoids compare?*

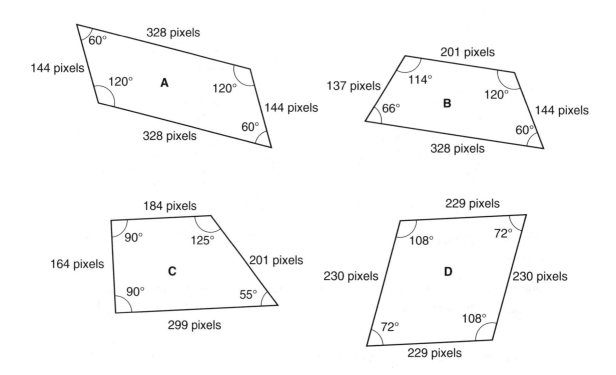

Additional suggestions for moving RH from Levels 2.2 to 2.3 are given in Chapter 3.

Chapter 3

Instructional Strategies for Geometric Shapes

Once you have used CBA Geometric Shapes assessment tasks to determine students' CBA levels of reasoning, you can use the teaching suggestions and instructional tasks described in this section to tailor instruction to precisely fit students' learning needs. For each major CBA level of reasoning, there are suggestions for teaching that encourage and support students' movement to the next important type of reasoning in the CBA sequence.

For students to make progress, have them do several problems of a specific type until you see them move to the next CBA level, or you become convinced that they are not quite ready to move on to the next level. In the latter case, try a different kind of problem suggested for that level.

Justifying Ideas

While justifying and proving ideas are critically important in all of mathematics, justification plays a particularly important role in the CBA levels for Geometric Shapes. Indeed, Level 4 reasoning focuses on formal geometric proofs. Because of the importance of justification in the CBA Geometric Shapes levels, it is appropriate in this chapter to make a general comment about the type of classroom environment that supports students' development of justification reasoning.

To promote the development of justification reasoning in the classroom, primary responsibility for establishing the validity, or "truth," of mathematical ideas should lie with students, not teachers or textbooks. In essence, each student is seen as a mathematician, somebody who is responsible for solving mathematical problems, making conjectures, and establishing the validity of his or her solutions and conjectures within the classroom culture. The teacher does not act as the mathematical authority. In fact, whenever students describe their mathematical problem-solving activity, they are implicitly claiming that their statements are valid. That is, they are justifying their mathematical ideas. Thus their explanations should be elaborate enough so that other students can evaluate their validity.

To help students' justification of mathematical ideas gradually develop and become increasingly more rigorous, instruction should focus on students building arguments to convince themselves and others of the validity of their ideas. It should permit students to utilize visual and empirical thinking because such thinking is the foundation for higher levels of geometric reasoning. It should involve students in the crucial elements of mathematical discovery and discourse—explanation, conjecturing, careful reasoning, and the building of justifying arguments that can be scrutinized by others. In this atmosphere, and with appropriate instructional tasks and teacher questioning, students will eventually see the limitations of visual and empirical attempts at validation and move toward logical deductive methods. Here are some examples of teacher questions that promote student justification of ideas:

How do you know that is true?

How can you convince your classmates that you are right?

Will that always be true?

Is your statement true for every rectangle?

Is what you said true for this triangle too, or just the triangle that you drew?

A Special Role for Computer-Based Dynamic Geometry

Working with dynamically manipulable geometric shapes on computer screens is intriguing and engaging to students. These dynamic geometry environments are extremely well researched, and the research indicates that use of such environments increases students' geometric knowledge and moves them to increasingly sophisticated methods of reasoning and justification (Battista, 2007a, 2007b, 2009; Battista and Clements, 1995; de Villiers, 2003; Hoyles and Jones, 1998; Laborde, 1998; Olive 1998). For these reasons, instructional use of dynamic geometry software such as *The Geometer's the Sketchpad*, *Cabri Geometry*, and *GeoGebra* (free online) is strongly recommended by the National Council of Teachers of Mathematics (NCTM, 2000). Because, if properly used, dynamic geometry is so engaging and effective, such software plays a major role in CBA suggestions for instruction.

What Is Dynamic Geometry?

Descriptions of the two basic uses of dynamic geometry software in elementary and middle school illustrates the nature of such software. In one use, appropriate for younger and older students, students explore draggable geometric constructions. A construction is draggable if it maintains its geometric properties when its vertices or sides are dragged with the mouse. An example is a computer Parallelogram Maker, which can be used to make any desired parallelogram that fits on the computer screen, no matter what its shape, size, or orientation—but only parallelograms. The appearance of a Parallelogram Maker is changed by dragging its vertices with the mouse (see below). Because of the way the Parallelogram Maker is constructed with geometric construction-based computer commands, no matter how this construction is manipulated, it remains a parallelogram.

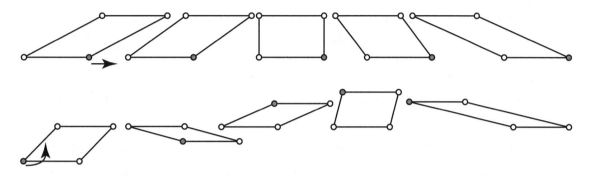

In the author's specially constructed Shape Makers dynamic geometry environment for *The Geometer's Sketchpad*, elementary and junior high school students explore draggable shapes without having to learn any of the construction tools of the software, making the activities accessible to students even in the primary grades. Draggable shapes, such as the Parallelogram Maker, are interesting, manipulable, visual-mechanical objects with movement constraints that can be conceptualized and geometrically analyzed (Battista, 1998, 2003). With proper instructional guidance, students' analyses of Shape Makers promote their construction of meaningful conceptualizations of geometric properties, which can be applied not only to draggable computer drawings but to the category of geometric objects these drawings generate.

The author's *Shape Makers* book consists not only of computer Shape Makers for each common type of quadrilateral and triangle but of a carefully designed sequence of instructional activities that research shows supports students' movement from CBA Level 1 to Level 3.4 (Battista, 2003, 2007a, 2007b, 2009; Borrow, 2000). In initial activities, students use the various Shape Makers to make pictures. These activities encourage students to become familiar with the movement possibilities of the Shape Makers viewed as holistic entities. Students then are involved in activities that require more careful analysis of shapes; they are guided to formulate and describe geometric properties of shapes. Unmeasured Shape Makers are replaced by Measured Shape Makers, which display angle measures and side lengths that are instantaneously updated when the Shape Makers are manipulated. Parallelism- and symmetry-testing capabilities are added. Finally, students focus on classification issues as they compare the sets of shapes that can be made by the various Shape Makers. (In addition to being available in the *Shape Makers* book, measured Quadrilateral and Triangle Makers created by the author are also available online at www.heinemann.com/products/E04351.aspx.)

In a second common use of dynamic geometry software (generally used in junior high school), students use geometric construction tools on the computer to create their own draggable shapes (this can be done in any of the dynamic geometry programs previously mentioned). This instructional use is discussed later in this chapter. However, it is important to recognize that it is fairly easy to create your own draggable Shape Makers in any of the dynamic geometry software programs. Once you have created these draggable Shape Makers, you can use any of the dynamic geometry instructional tasks suggested in the chapter.

As an example, consider how to make a Parallelogram Maker in *The Geometer's Sketchpad*.

Step 1. Construct two line segments that intersect at their one pair of endpoints.

Step 2. Construct a line parallel to side AB, through point C, and a line parallel to side BC, through point A. (There is a computer command for constructing parallels.)

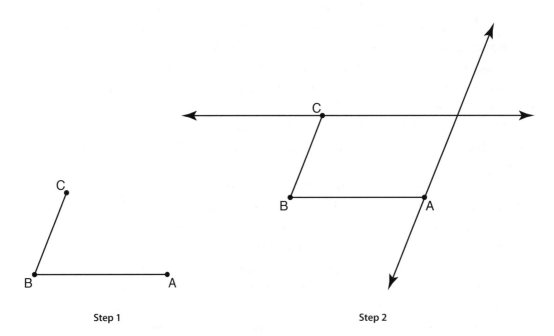

Step 3. Construct the intersection point of these two lines, point D, hide the lines, then draw line segments AD and CD. We now have a draggable Parallelogram Maker, ABCD. (Our construction explicitly created opposite sides parallel.)

Step 4. Use computer commands to measure the side lengths and angles of ABCD. These measurements automatically change as you manipulate this Measured Parallelogram Maker. (Length measurements below are in pixels, which are the tiny "dots" that make up pictures on the computer screen. Pixels are used instead of centimeters to avoid decimals, but you can choose to use centimeters and decimals with older students.)

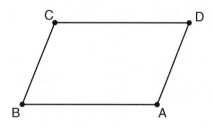

Step 3

Length (BA) = 303 pixels   Angle A = 112°
Length (BC) = 196 pixels   Angle B = 68°
Length (CD) = 303 pixels   Angle C = 112°
Length (AD) = 196 pixels   Angle D = 68°

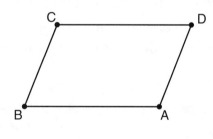

Step 4

Instructional Strategies for Geometric Shapes

Teaching Students at Level 1.1: Helping Students Correctly Identify Shapes

To help students progress from incorrect identification of shapes (Level 1.1) to correct identification (Level 1.2), they should be put in situations in which they are given feedback on their shape identifications. For instance, students can play the Guess My Rule game. (This game also encourages students to reflect on their shape identifications, which encourages parts-based reasoning.)

Guess My Rule Game

Goal: To increase students' ability to discriminate shapes, to start attending to the parts of shapes, and to see commonalities among examples in a group of shapes.

Directions: Make copies of the shapes shown in Groups 1 and 2 on the following pages. (Use overhead transparencies or sheets of paper for a document camera, or premade images for a smartboard—do not draw them because freehand drawing is too inaccurate.) Cut out individual shapes and put Group 1 and Group 2 shapes in separate envelopes.

Tell students:

I'm thinking about a special group of shapes.

There is a rule for belonging to the group.

Your job is to figure out the rule.

As I show you shapes, I will tell you if they belong to the group or not.

Show the first two shapes in Group 1 (Shapes A and B), placing them at the top of the overhead screen. (Place all shapes so that the letters on the shapes are right-side up.)

Say, "*These two shapes belong to the special group. For each shape that I show you, put a thumb up if you think the shape belongs to the group, and a thumb down if you think the shape does not belong to the group. I will then put the shape at the top of the screen if the shape belongs to the group, and at the bottom if it does not.*"

After you have shown all shapes, one at a time, ask students what the rule is for belonging to the group. (A shape belongs to Group 1 if it is a triangle. A shape belongs to Group 2 if it is a rectangle.)

CBA Guess My Rule Activity: Group 1

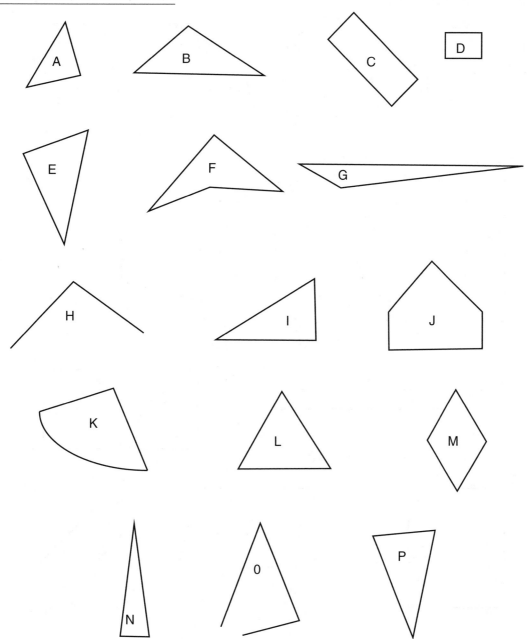

CBA Guess My Rule Activity: Group 2

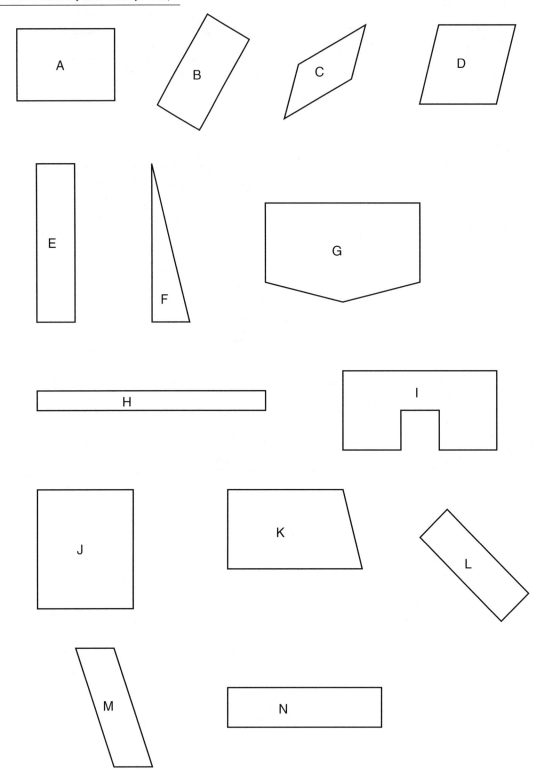

Generally, students will describe the rule by shape identification. For example, for Group 1, students will probably say that the shapes are triangles. If students say something like, the shapes have 3 points, or the shapes have 3 sides, you might ask them, *What is a shape with 3 sides and points called?*

However, even if students use the classification "triangle" for Group 1 (or "rectangle" for Group 2), don't assume that they know exactly what these shapes are. Ask, *"Why does this shape belong in Group 1? Why do you say it is a triangle? Why does this shape not belong in Group 1? Why is it not a triangle?"*

Teaching Students at Level 1.2: Helping Students Move Toward Parts- and Property-Based Reasoning

To help students progress from visual-holistic reasoning about shapes (Level 1) to parts- and properties-based reasoning (Level 2), they should be put in situations in which they have to *justify* their shape identifications and more carefully analyze shapes.

For instance, as a class, students can be shown the shapes below and asked, "Which shapes are squares? How do you know? How can you convince somebody else your answer is correct?" The disagreements that students have about which shapes are squares and *why* they are squares will encourage students to reflect more deeply about what squares are. This reflection will lead students toward understanding squares in terms of their *parts* and how these parts are related.

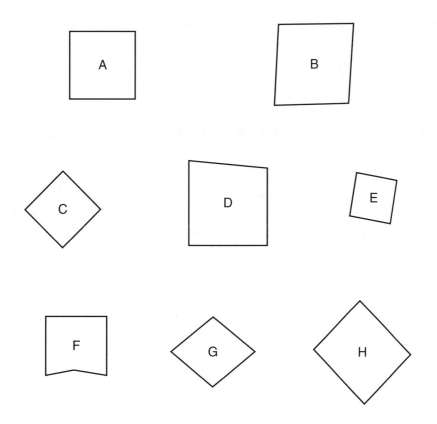

Instructional Strategies for Geometric Shapes

Similarly the CBA Drawing Machine Assessment Tasks (Problems 1–3 in the Appendix) can be used as *instructional* activities to encourage and support students' transition from the holistic thinking of Level 1 to the parts-and-properties reasoning of Level 2. As shown below, such tasks, along with proper teacher questioning, can encourage students to move toward property-based thinking.

These are shapes made by Drawing Machine 1. How are these shapes alike? What's the rule?

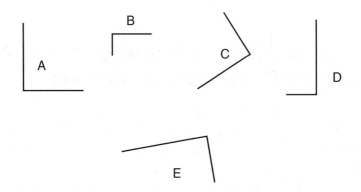

Student: *They are L's.*

Teacher: *What's the rule?*

Student: *The shapes are L's?*

Teacher: *What makes them L's?*

Student: *They look like square corners; the lines come together like straight.*

By asking an appropriate probing question, "What makes them L's?", the teacher encourages the student to move from her holistic Level 1 thinking (they are L's) to Level 2.1 reasoning about *parts* of shapes (the lines come together like straight).

Using Computer Shape Makers or CBA Applets to Encourage Property-Based Reasoning

To help students progress from holistic identification of shapes toward property-based identification (and more accurate identification), you can use a dynamic geometry Rectangle Maker. After demonstrating and explaining to students that the computer Rectangle Maker makes rectangles—and having students manipulate it—ask them to predict which of Shapes A–G the machine can make in the figure below.

Move vertices on the Rectangle Maker to see the kinds of shapes it can make. Predict which of Shapes A–G the Rectangle Maker can make. Check your answers with the Rectangle Maker.

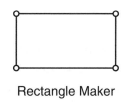

Rectangle Maker

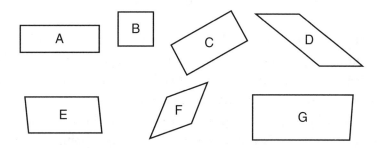

Students check their answers by trying to manipulate the Rectangle Maker to exactly overlap the shapes. Asking students *why* shapes can or cannot be made by the Rectangle Maker encourages them to focus on the structure of the shapes, which can best be understood by attending to the *parts and properties* of the shapes.

Teaching Students at Levels 2.1 and 2.2: Helping Students Understand and Use Formal Geometric Concepts to Analyze Shape Properties

For students to progress from informal descriptions of shape properties (Levels 2.1 and 2.2) to complete and correct formal descriptions (Level 2.3), they must develop understanding of two interrelated ideas: (a) property-based characteristics of shapes; and (b) formal geometric concepts that are used to describe shape properties. For example, for a rectangle, students must first notice that the adjacent sides of a rectangle meet in a special way, then they must describe this special relationship between adjacent sides by using the formal geometric concepts of right angle, perpendicularity, or 90° angles. So students must not only participate in rich discussions of properties of shapes, they must also discuss critical concepts such as the length of line segments, the measurement of angles, parallelism, and right angles.

Building Understanding of Formal Concepts for Analyzing Geometric Shapes

A number of geometric concepts are commonly used to analyze 2-dimensional (2D) shapes. Although it is beyond the scope of this book to discuss instruction for each of these concepts, this section illustrates the type of instruction that can help students develop appropriate understanding of some of them. Achieving Level 2.3 requires that students understand the formal concepts described in this section.

Paths and Polygons

Help students think about 2D shapes as paths. A *path*, or continuous path, is what we get when we continuously move a point through the plane. For instance, we get a path when we drag a pencil on a piece of paper, without lifting it up at any time during this motion. You can help students think about paths by asking them to trace a figure that you give them without lifting their pencil or going over any part of the figure more than once.

One special kind of path is a *simple closed path*, which starts and ends at the same point (*closed*) and does not intersect itself anywhere except its start/end point (*simple*). A circle and square are simple closed paths. Two basic kinds of shapes typically are studied in elementary school geometry—simple closed paths with curved parts, like a circle or ellipse, and polygons. A *polygon* is a simple closed path consisting of line segments only (a *line segment* is a straight path between two points). Other geometric concepts that are important in analyzing 2D shapes are *points*, *lines*, *angles*, *symmetry*, and *congruence*, which are all defined in the Glossary.

To help students understand many of these geometric concepts, describe or define the concept, then give students an opportunity to identify examples and nonexamples of the concept and justify their answers. For instance, you could explain to students that a *polygon* is a simple closed path consisting of line segments only, and then have students decide which of the figures below are polygons and justify their answers. (Of course, prior instruction would have to deal with the concepts of paths, simple and closed paths, lines, line segments, and straightness.)

Measurement

Numerous formal geometric properties are described using length and angle measurement. For instance, one critical property of rectangles is that opposite sides are equal in length measurement. Another property is that all the interior angle measurements must be 90°. Before students can fully understand formal measurement-based properties, we must be sure that they understand length and angle measurement.

Understanding Length Measurement

The length of a line segment is measured by determining how many length-units it takes to completely cover the segment with no gaps or overlaps. For instance, to say that the black segment below is 5 centimeters long means that it takes 5 one-centimeter length-units to cover the segment, end to end.

See the CBA book *Cognition-Based Assessment and Teaching of Geometric Measurement* for a detailed discussion of instructional activities that can help students understand length measurement. To genuinely understand the length measurement ideas used in specifying geometric properties, students should be at least at Level M4 in the CBA length learning progression. See **STUDENT SHEETS 1 AND 2** for activities that can help you check and review the concept of length. (All of the student sheets referenced in this chapter can be found at www.heinemann.com/products/E04351.aspx. Click on the "Companion Resources" tab.)

Understanding Angle Measurement[1]

Angles are measured in degrees. One degree is the amount of rotation it takes to move the bottom ray onto the top ray as shown below.

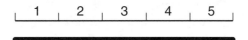

It takes 360 one-degree (1°) rotations to make one complete revolution, all the way around a circle.

It takes 10 one-degree rotations to make a 10° angle.

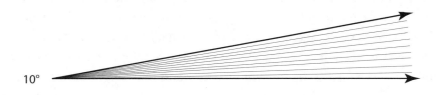

---

[1] See Battista, 2003, for more suggestions for teaching angle measurement.

Instructional Strategies for Geometric Shapes

The degree measure of an angle tells us how much rotation is required to turn one side of the angle onto the other. But there is one troublesome point that arises in angle measure. When 2 rays intersect at their endpoints, there are two possible rotations that move one ray onto the other, so there are two possible angle measures. For instance, both measurements of the following angle make sense:

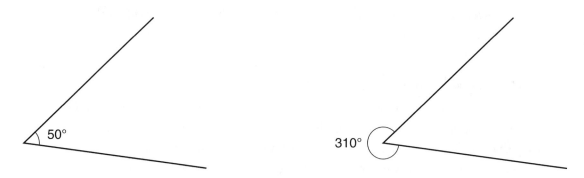

Some textbooks deal with this issue by always choosing the measure that is 180° or less. Other textbooks deal with the issue by indicating the starting and ending position of the rotation of rays in the angle, as shown above. You can use whichever treatment is consistent with your curriculum, but eventually (in trigonometry) students will have to learn the latter treatment.

Instructional Tasks on Degree Measure of Angles

After explaining degrees as shown above, ask students the following questions:

- First get students familiar with full, half, and quarter turns. Say, "Everyone stand up and face the front of the room. If we turn all the way around so that we are facing the front of the room again, that is 1 full turn. Suppose we turn half way around, where would we be facing [*the back of the room*]? How much do we have to turn to be facing the front [*half turn*]? Now everyone face the front of the room. Suppose I asked you to make a quarter turn, or a fourth of a turn, to your right. Where would you be facing? How about another quarter turn to the right? If I face the front, how many quarter turns to the right do I need to be facing the left wall. You can give various sequences of commands for students to follow (they will check with each other): Face front. Make 1 full turn (either way). Make a quarter turn to the left. Make a half turn to the left. Which wall are you facing? After students have done this for a while, you can even have them close their eyes while making the turns.

- Once students seem to physically understand turns, help them relate turns to angle measurements. "How many degrees did we say were in a whole turn?" (360) Have students stand up and turn a half turn. Ask, "How many degrees are in a half turn?" (180) Do the same for a quarter turn (90), and a three-quarter turn (270). For younger students, you might even make a 360° protractor that a student can stand on as she turns. (Mark multiples of 10°.)

- What does "doing a 360" mean?
- You can play "Guess the angle" games with students. Show an angle, have students estimate its measure, then show them the actual measurement. (Computer versions of such games can also be found on the Internet.)
- How many 10-degree angles must one turn through to make a full revolution?
- How many degrees does the hour hand on a clock rotate through in 1 hour?
- How many degrees does the second hand on a clock rotate through in 1 second?

Another good activity is to have students estimate the measure of drawn angles, then measure them with a protractor (you could also make an angle measuring tool in dynamic geometry).

Parallelism

Two lines in a plane are parallel if they do not intersect. Two line segments in a plane are parallel if the lines through them are parallel. Again, you can give students examples and non-examples of parallelism and have them decide and justify their answers (see below). Note that some students think that to be parallel, two line segments must be parallel and equal in length. For pairs C through E, be sure to ask students to imagine the two lines that contain the line segments—we judge whether segments are parallel by examining whether the lines through them are parallel.

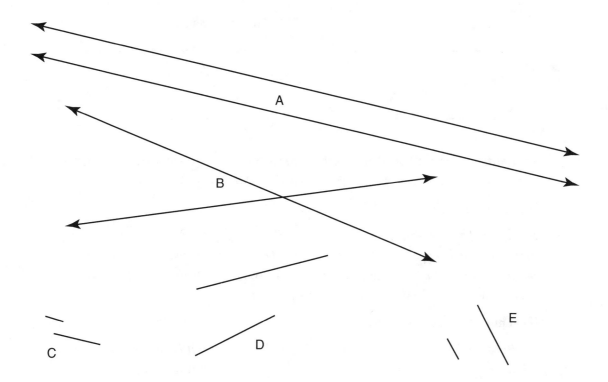

Instructional Strategies for Geometric Shapes

Moving from Informal Descriptions of Shape Properties (Level 2.1) to Formal Descriptions (Levels 2.2 and 2.3)

To help students move from informal and vague descriptions of shape properties to complete and correct formal descriptions (from Level 2.1 to Level 2.2 and into Level 2.3), students can be given paper-and-pencil tasks that require them to find common characteristics for a group of shapes and make careful distinctions between shapes, such as Problem 2 in the Appendix (page 115). Keep in mind that any of the CBA assessment tasks also can be used as instructional tasks.

The episode below illustrates how using computer-based dynamic geometry tasks, along with appropriate teacher questioning, can encourage students to move from holistic identification to correct property-based characterization of shapes (Battista, 2003). During a dynamic geometry computer activity, a student discusses her attempts to get a Rhombus Maker to make Shapes X and Y. Asking students *why* shapes can or cannot be made by such a tool encourages them to focus on the structure of the shapes, which can best be understood by attending to the *parts and properties* of the shapes.

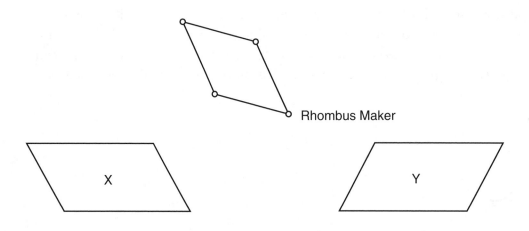

Student: *I think the Rhombus Maker can make Shape X.* [After trying it] *It doesn't work. I think I might have to change the Rhombus Maker to make Shape Y.*

Teacher: *Why Y?*

Student: *The Rhombus Maker is like leaning to the right. On X, the shape is leaning to the left. I couldn't get the Rhombus Maker to lean to the left, and Y leans to the right, so I'm going to try it.* [After her initial attempts to get the Rhombus Maker to fit exactly on Shape Y] *I don't think that is going to work.*

Teacher: *Why are you thinking that?*

Student: *When I try to fit it on the shape, and I try to make it bigger or smaller, the whole thing moves. It will never get exactly the right size.* [Manipulating the Rhombus Maker] *Let's see if I can make the square with this. Here's a square.*

Teacher: *You said the Rhombus Maker could make the same shape as Shape Y, what do you mean by that?*

Student: *It could make the same shape. It could make this shape, the one with 2 diagonal sides and 2 straight sides that are parallel. It could have been almost that shape, and it got so close I thought it was that shape. [Continuing to manipulate the Rhombus Maker] Oh, I see why it didn't work, because the 4 sides are even, and Shape Y is more of a rectangle.*

Teacher: *How did you figure that out?*

Student: *Well, I was just thinking about it. If the Rhombus Maker was the same shape, then there is no reason it couldn't fit on Y. But I saw when I was playing with it to see how you could move it and things like that, that whenever I made it bigger or smaller, it was always like a square, but sometimes it would be leaning up, but the sides are always equal.*

This episode clearly illustrates how manipulation of a dynamic geometry Rhombus Maker *and reflection on that manipulation*, encouraged by the teacher, can enable a student to move from thinking holistically to thinking about interrelationships between a shape's parts—that is, about its geometric properties. The student began the episode thinking about the Rhombus Maker and shapes holistically and vaguely, saying that she was trying to make the Rhombus Maker "lean to the right," and get "bigger or smaller," and that "the whole thing moves." The fact that she could not make nonequilateral parallelograms [Shapes X and Y] with the Rhombus Maker caused her to reevaluate her conception. As she continued to analyze why the tool would not make the parallelogram—why it would not elongate—her attention shifted to its side lengths. This new focus of attention enabled her to see that the sides of the shapes from the Rhombus Maker were the same length. As she incorporated this understanding about side lengths into her conception of the Rhombus Maker, she developed a property-based (Level 2.3) understanding of the Rhombus Maker and why the Rhombus Maker could not make Shape Y. She also developed an understanding of the concept of rhombuses and why Shapes X and Y were not rhombuses.

You can use "measured" quadrilateral and triangle makers to encourage more sophisticated Level 2.3 reasoning. See the Measured Rectangle Maker below. Because these measured computer Shape Makers show how the measures of sides and angles change as the shape changes, they can help focus students' attention on properties that involve side and angle measurements.

Length (AB) = 100 pixels   Angle (A) = 90°
Length (BC) = 140 pixels   Angle (B) = 90°
Length (CD) = 100 pixels   Angle (C) = 90°
Length (DA) = 140 pixels   Angle (D) = 90°

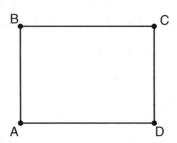

For example, as students manipulate a Measured Rectangle Maker on a computer, ask questions such as, "In what ways are all rectangles the same? In what ways can they be different? What do the length and angle measurements of the Rectangle Maker tell you about the properties of rectangles? Predict which shapes can be made by the Rectangle Maker (e.g., see **STUDENT SHEET 4**, Part 2 ⬇), then check your predictions on the computer." You can do similar activities with other Measured Shape Makers.

Although somewhat less effective and engaging for students, for rectangles you can use activities like those on **STUDENT SHEET 4**, Parts 1 ⬇ and 2 to engage wstudents in similar reasoning.

Questioning Is Key

Asking probing questions can be critically important in encouraging students to use more sophisticated descriptions of shape properties. For instance, a student was working on the task below.

These are shapes made by Drawing Machine 1. How are the shapes alike? What's the rule?

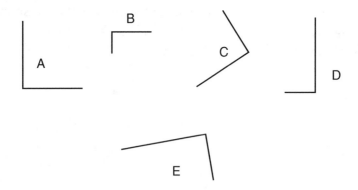

Student: *The shape machine makes square corners, like in a square or a rectangle.*

Teacher: *What makes them square corners?*

Student: *They are 90° angles.*

In this case, the teacher's probing question encouraged the student to rethink her informal description of a property and give a formal description. The transition from informal to formal descriptions does not occur naturally for most students; it must be encouraged and nurtured by teachers. However, as this example illustrates, for many students this transition may only require well-chosen and well-timed questions.

Advanced Guess-My-Rule Problem

More advanced Guess-My-Rule problems, such as the one below, can be used to encourage Level 2.3 reasoning.

For the two activity pages that follow, make one paper copy and one copy from overhead transparencies (or use a document camera or smartboard). Cut out the shapes on the transparency page, leaving behind the y's and n's (*y* means the shape is an example of the special group; *n* means the shape is *not* an example of the special group). The paper copy is for your reference only (so you can recall which shapes are examples and nonexamples).

I'm thinking about a special group of shapes.

There is a rule for belonging to the group.

Your job is to figure out the rule.

I will tell you if shapes belong to the group or not.

Show the first two shapes, placing them at the top of the display. (Place all shapes so that the numbers on the shapes are right-side up.)

Say, "*These two shapes belong to the special group. For each shape that I show you, put a thumb up if you think the shape belongs to the group, and a thumb down if you think the shape does not belong to the group. I will then put the shape at the top if the shape belongs to the group, and at the bottom if it does not.*"

After you have shown all shapes, one at a time, ask students what the rule is for belonging to the group.

Answer: A shape is a member of the group if it is a quadrilateral that has 2 pairs of congruent sides. An alternate rule is, A shape is a member of the group if it is either a parallelogram or a kite (or both). (*Note:* This particular example is for older students; they should be at least at Level 2.3. Similar examples can be done with younger students, but without measurements, and targeting a simple shape class. For instance, you could have students distinguish triangles or quadrilaterals from other polygons.)

Once students have described both rules, ask them why either rule works. (In a parallelogram, the 2 pairs of opposite sides are congruent. In a kite, 2 pairs of adjacent sides are congruent.)

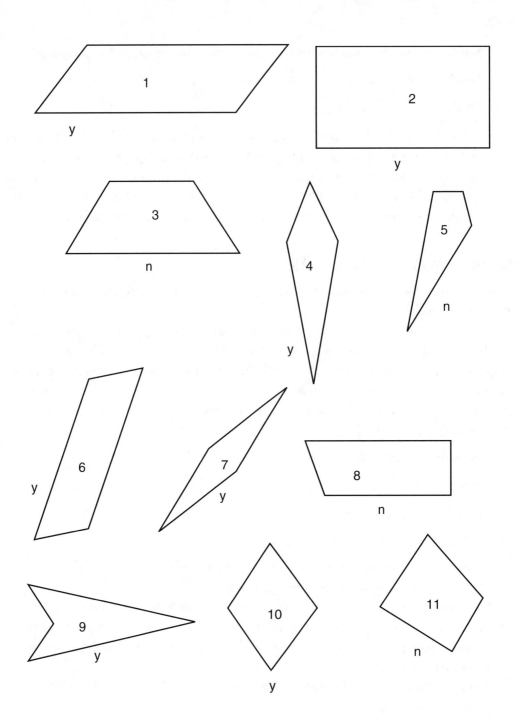

CBA Activity Sheet

Length (AB) = 66 pixels   Angle (A) = 68°
Length (BC) = 97 pixels   Angle (B) = 124°
Length (CD) = 97 pixels   Angle (C) = 44°
Length (DA) = 66 pixels   Angle (D) = 124°

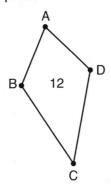

y

Length (AB) = 89 pixels   Angle (A) = 117°
Length (BC) = 117 pixels   Angle (B) = 63°
Length (CD) = 89 pixels   Angle (C) = 117°
Length (DA) = 117 pixels   Angle (D) = 63°

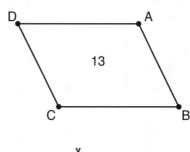

y

Length (AB) = 102 pixels   Angle (A) = 120°
Length (BC) = 73 pixels   Angle (B) = 120°
Length (CD) = 175 pixels   Angle (C) = 60°
Length (DA) = 73 pixels   Angle (D) = 60°

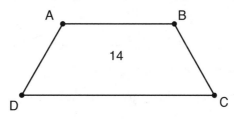

n

Instructional Strategies for Geometric Shapes   85

Another Illustration of Using Computer Shape Makers to Encourage Students to Move from Informal to Formal Property-Based Reasoning

Task: *Students are to determine which of Shapes 1–7 can be made by a computer Rectangle Maker* (Battista, 2003).

Working together, students M and T predicted that the Rectangle Maker could make Shapes 1–3, but not 4–6. They are now checking and discussing their results. Pay close attention to how the teacher encourages students to reflect more deeply on their informal conceptions of properties of rectangles.

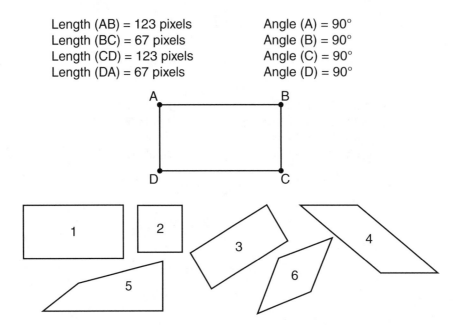

After using the Rectangle Maker to check Shapes 1–3, M and T move on to Shape 4.

T: *I'm positive the Rectangle Maker can't make Shape 4.*

M: *The Rectangle Maker has no slants. We had enough experience with Shape 3, that it can't make a slant.*

T: *Yes it can, it has a slant in Shape 3, and we made Shape 3 with the Rectangle Maker.*

Teacher: *What do you mean by slant?*

M: *Like this. See how this is shaped like a parallelogram* [motioning along the perimeter of Shape 4].

T: *This is in a slant right now* [points at the Rectangle Maker, which is rotated from the horizontal].

Note that "slant" means nonperpendicular sides for M, but rotated from the horizontal for T. . . .

> **M:** *It can't make a shape like Shape 4. . . . It can't make something that has a slant at the top and stuff.*
>
> **T:** *Do you mean it has to have a* straight *line right here* [pointing to Shape 6], *like coming across?* [T now uses "straight" to mean horizontal.]
>
> **M:** *I know they are* straight, *but they are at a slant and the Rectangle Maker always has lines that aren't at a slant. . . .*
>
> **Teacher:** [After several failed attempts to unobtrusively get students to reconceptualize the "slant" in terms of angles] *Keep on really looking at what makes these different. And maybe some of the information on the screen will help you. Watch the numbers up there and see if that will help you. . . .*
>
> **T:** [Manipulating the Rectangle Maker after the teacher leaves] *Oh, the Rectangle Maker always has to be a 90° angle. And Shape 4 does not have 90° angles. And so Shape 3 has to have a 90° angle, because we made this with the Rectangle Maker. So there is one thing different. A 90° angle is a right angle, and Shape 4 does not have any right angles.*

M and T were trying to find a way to conceptualize and describe the spatial relationship that sophisticated users of geometry describe by saying that the sides of the Rectangle Maker are perpendicular. In so doing, they used familiar terminology and concepts like "slanted" and "straight" that inadequately described the idea with which they were grappling.

While interacting with M and T, the teacher recognized that they were unable to conceptualize the situation in terms of the formal geometric concepts of perpendicularity or right angles. Though she asked numerous questions that she thought might activate a formal description, M and T were unable to think of the situation in formal terms.

However, after the teacher left the boys, T manipulated the Rectangle Maker, focusing on its measurements. Through this manipulation, he discovered and abstracted that the Rectangle Maker always has four 90° or right angles. Furthermore, he abstracted this property sufficiently so that he was able to use it to analyze the differences between Shapes 3 and 4. He subsequently conceptualized that the Rectangle Maker could not make Shapes 4, 5, and 6 because they do not have 4 right angles. In fact, by the end of the class period, the boys saw that the spatial relationship to which they were attending could be described in terms of the formal mathematical concept of right angle.

This episode illustrates how difficult it can be for students to reformulate unrefined informal geometric ideas in terms of formal geometric concepts. It is only after much guidance, reflection, and experimentation that students can meaningfully use formal geometric conceptualizations.

Further Encouraging the Transition to Complete and Correct Formal Descriptions of Properties (Level 2.3)

Hinged Rods

Give students two pairs of equal hinged rods (such as in Lego Technics building sets) as shown below. (You could also use popsicle sticks with holes at the end, connected by metal brads.)

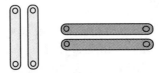

Ask students what kinds of quadrilaterals they can make from the rods, and have them explain why their answers are correct. Students should discover that they can make either parallelograms or kites, as shown below.

Parallelogram　　　　　　　　　　　**Kite**

Help students characterize these two possibilities using formal geometric properties:

- If the equal sides are across from each other (opposite), the quadrilateral will be a parallelogram.
- If the equal sides are next to each other (adjacent), the quadrilateral will be a kite.

Note that if all sides are equal, you get a rhombus, which is both a parallelogram and a kite.

Questioning

In the following examples, students' descriptions start out either informal or incomplete (that is, below Level 2.3), but jump up to Level 2.3 because of both the demands of the tasks and appropriate teacher questioning. These examples give further guidance about how to encourage students to move to more sophisticated reasoning.

How are these shapes alike? What's the rule?

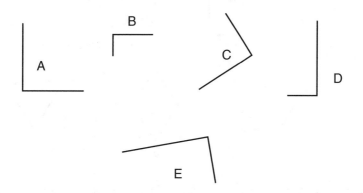

Student: *The shape machine makes L's. And it's just L's flipped around or sideways or upside down. They are alike, like the letter L.*

Teacher: *What's the rule?*

Student: *The shapes are an L.*

Teacher: *What makes them an L?*

Student: *The 90° angle that they are at.*

Teacher: *What do you mean by that?*

Student: *Like 90° angles are straight up and across, and the L is straight up and across. And it's an L at a 90° angle.*

It is not until the teacher asks, "What makes them an L?" that the student uses the formal property "90° angles." If students still have difficulty, ask "What are the parts of the L's *[2 line segments and an angle]*? Is there anything special about these parts? How are the parts of A like the parts of B? How do the angles in the shapes compare?"

Instructional Strategies for Geometric Shapes

How are these shapes [A–G] alike? What's the rule?

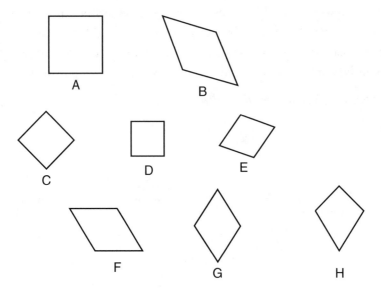

Student: *They all have 4 lines and 4 slants. I mean not 4 slants, 4 points. But that might not be it. They all kind of look like diamonds.*

Teacher: *This shape (H) cannot be made by this drawing machine. Why?*

Student: *Oh, these have the same length on each side* [tracing over the sides on Shape B].

Teacher: *So what do you think the rule is for the drawing machine?*

Student: *All the sides have the same length.*

Teacher: *Which of these shapes can be made by the drawing machine? Why or why not?*

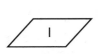

Student: [Student circles Shapes J and L]. *This, it can't make it* [circling Shape K].

Teacher: *Why can't it?*

Student: *Because it doesn't have the same lengths on each side.*

Teacher: *What about Shape I, why can't it make it?*

Student: *Because that one doesn't have the same length on each side.*

Teacher: *And how about Shapes J and L, why can it make those?*

Student: *Because they have the same size on each side.*

It's important to note that this student started his analysis at Level 2.1, using informal descriptions of properties (lines, slants, and points). Only when he is shown a shape that the drawing machine cannot make and asked *why* it cannot make this shape does he refine his thinking to precisely describe a formal property of the shapes—same length on all sides. Thus, asking students to do more than describe a set of examples—to discriminate examples from nonexamples and to justify their discriminations—can encourage them to refine their thinking and progress to a higher level of reasoning.

Note that the Level 2.3 reasoning the student used could have been sharpened if the problem had shown the side measures for the shapes.

These shapes were all made with Drawing Machine 4. How are the shapes alike?

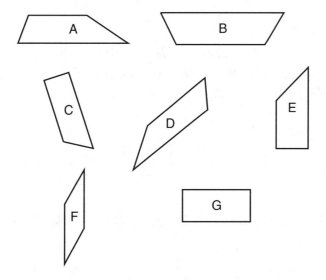

Student: *They're going to be rectangles but then they're cut off* [motions to the end of D]. *They all have 4 sides.*

Teacher: *This shape cannot be made by Drawing Machine 4. Why?*

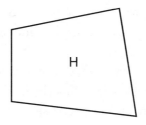

Instructional Strategies for Geometric Shapes

Student: *Drawing Machine 4 draws shapes that . . . each shape has at least one set of sides that are parallel to each other.* [Draws marks on the figures to indicate which sides are parallel.]

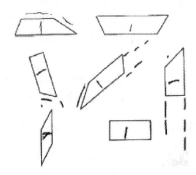

Student: *This one [H] doesn't have parallel sides.* [Draws how the top and bottom sides of H would intersect if they were continued to the left by extending the sides out as dotted lines.]

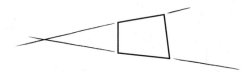

Teacher: *Which of these shapes can be made by Drawing Machine 4? Why or why not?*

[Student circles I and J. He indicates how the top and bottom of each, when extended as dotted lines, run parallel to each other.]

As in the previous example, when this student was shown a shape that the drawing machine could not make, he refined his thinking to provide a precise formal property of the examples.

Another type of task that can encourage complete and correct formal property-based reasoning (Level 2.3) are tasks like those in Assessment Problems 6 and 7 (Appendix, pp. 123 and 124) (see also **STUDENT SHEET 3**). In such tasks, to answer correctly, students must attend to precise *measurement*-based properties. Visual-holistic or informal property-based reasoning generally produces wrong answers. For instance, consider parts (a) and (b) of Problem 7 in the assessment tasks.

The measurements for a shape are given below.

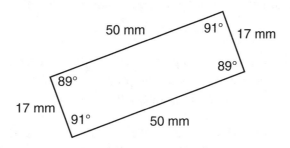

For each statement about this shape, circle T if the statement is True, or F if the statement is False. If you can't tell if the statement is true or false, circle Can't tell.

For each statement, describe how you would convince someone that your answer is correct.

(a) The shape is a square.   T   F   Can't tell

(b) The shape is a rectangle.   T   F   Can't tell

Teacher: [Whole class discussion, after students have worked in small groups on the task] *OK, let's talk about what you think. I'd like some of you to tell us what your answers are and explain why you think your answers are correct.*

Student 1:
(a) *No, it's too long to be a square.*
(b) *Yes, it looks like a rectangle.*

Student 2:
(a) *No, these sides* [top and bottom] *are too long to be a square.*
(b) *Yes, it has 2 long sides and 2 short sides.*

Student 3:
(a) *No, the sides are not equal like they are in squares.*
(b) *Yes, because the sides across from each other are equal.*

Student 4:
(a) *No, the sides are not equal, and they have to be in a square.*
(b) *No, because this shape doesn't have four 90° angles.*

Teacher: *OK, we seem to be in agreement about whether this shape is a square. You all seem to think it is not a square. Why is it not a square? Student 3?*

Student 3: *Well, in a square the sides are equal. These sides are not equal. Besides it doesn't even look like a square.*

Teacher: *What do you think Student 2?*

Student 2: *Yeah, I agree. The top and bottom sides are longer than the right and left sides, so the sides are not equal.*

Teacher: *Student 1?*

Instructional Strategies for Geometric Shapes

Student 1: *I agree, because it's too long, its sides aren't equal.*

Teacher: *So what property of squares does this shape not have?*

Student 2: *That all its sides are equal.*

The teacher could have asked at this point what the other properties of squares are, but she decided it was better to focus on the problems from the student sheet that the students had already thought about in small groups.

Teacher: *Alright, so we seem to agree that the shape is not a square because its sides are not all the same length. Now, what about, Is it a rectangle? We seem to have some disagreement. Student 2?*

Student 2: *It looks just like a rectangle, and it has 2 long sides and 2 short sides.*

Student 3: *I agree.*

Student 4: *But it doesn't have 90° angles.*

Teacher: *Why does a rectangle have to have 90° angles?*

Student 4: *Because if it doesn't, then it will be tilted or slanted like this [draws].*

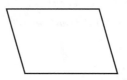

See, these angles are not 90° and the sides are slanted. It's not a rectangle. In this shape [on student sheet], the sides are slanted so little you can't really see it. But they are slanted so it is not a rectangle.

Student 3: *I guess that makes sense. When we used the [computer] Rectangle Maker, it always had 90° angles, and it would not make slanted shapes like Student 4 drew.*

Teacher: *What do you all think?*

Student 2: *Maybe, but it still looks like a rectangle to me.*

Teacher: *OK, great discussion. Let's use our computer Rectangle Maker and see exactly what kinds of shapes it can make. Can it make the shape on our student sheet? I would like you to all try it, then we'll talk some more.*

Note first that students demonstrated a variety of levels of reasoning on this task, from Level 1 through Level 2.3. Note second, that through questioning, the teacher exposed all students to this variety of reasoning. For squares, she tried to get students to all see that the shape is not a square because one of the properties of squares is violated. By having Students 1 and 2 comment after Student 3 claims that a square's sides must be equal, she was giving them a chance to relate their informal characterizations to the more formal language used by Student 3. For emphasis, she repeats the formal, property-based reason for the shape not being a square in her one-sentence conclusion of the discussion.

For the question about rectangles, the teacher tried to move the disagreeing students toward a consensus. She wanted to capitalize on Student 4's correct reasoning. But she did not validate this reasoning by saying it was correct. She let the students attempt to validate it. She is trying to move the students toward justifying their mathematical conclusions with reasoning rather than reference to answer-giving authorities (of course, sometimes we need to reference authority). Her question to Student 4 allowed him to explain why he thought his reasoning was correct—to explain his reasoning in a way that was likely comprehensible to most students in the class. But she recognized that some students still needed additional exploration to make sense of Student 4's correct ideas.

Quadrilateral Riddles

Riddles that focus on properties can also be used to encourage genuine understanding of formal property-based reasoning (Level 2.3). Students choose from the following shapes: square, rhombus, rectangle, parallelogram, kite, trapezoid, quadrilateral. One of the most accessible ways for students to make progress on this type of task is using dynamic geometry Shape Makers for each of the 7 types of quadrilaterals. However, this task is thought-provoking even without computers. Once students have developed some notion of the definitions for various quadrilaterals, questions like those below can help them fully understand the various properties of these quadrilaterals. (Almost no students develop this deep understanding merely by reading definitions and seeing a couple of examples.)

1. I always have at least 2 lines of symmetry. Sometimes not all of my sides are equal. Which quadrilateral am I?

2. I can have 2 right angles or 4, but never just 1. Which quadrilateral am I?

3. I always have 4 pairs of supplementary angles. Sometimes I have some unequal sides. I don't always have right angles. Which quadrilateral am I?

4. I have side lengths 100mm, 100mm, 150mm, 150mm. Which quadrilaterals could I be?

5. I have angles 20°, 160°, 20°, 160°. Which quadrilaterals could I be? (Battista, 2003)

You can help students with these problems by asking appropriate questions. For instance, for Problem 1, ask: Which of the types of shapes that we have studied always has at least 2 lines of symmetry? How many lines of symmetry must a square have? How about a kite? OK, a square always has 4 lines of symmetry. Can a square be the mystery quadrilateral? Why not? Do lines of symmetry have to pass through vertices? Let's think about the lines of symmetry for a Kite Maker, Rectangle Maker, and Rhombus Maker (see below; students can use drawing, paper folding, or reflections in computer dynamic geometry). Now what do you think?

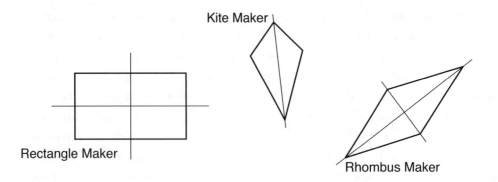

Teaching Students at Level 2.3: Moving to Empirically Relating Properties

Investigating Other Properties

Once students have reached Level 2.3, there are other shape properties that students can and should investigate. These properties are best investigated empirically.

For example, students can explore dynamic geometry Shape Makers for quadrilaterals and triangles to discover that the sum of the interior angles of a triangle is 180°, and for a quadrilateral 360°.

Triangle Maker

Length (AB) = 189 pixels   Angle (A) = 104°
Length (BC) = 135 pixels   Angle (B) = 44°
Length (CA) = 104 pixels   Angle (C) = 32°

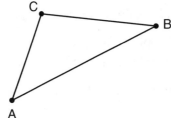

Quadrilateral Maker

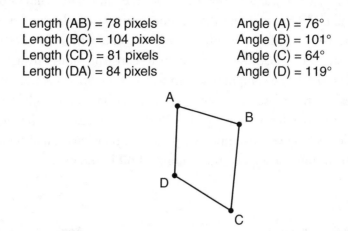

Length (AB) = 78 pixels Angle (A) = 76°
Length (BC) = 104 pixels Angle (B) = 101°
Length (CD) = 81 pixels Angle (C) = 64°
Length (DA) = 84 pixels Angle (D) = 119°

The justifications that students can make sense of at this level are intuitive. For instance, to show that the sum of the interior angles of a triangle is 180°, you can have each student cut out a large paper triangle. The students can cut the 3 "angle" sections (A, B, C) off their triangles and place them next to each other to form a straight angle, which measures 180°.

Students can do a similar thing with quadrilateral angles, but keep in mind that they must look at the interior angles (one of which will be greater than 180° for concave quadrilaterals).

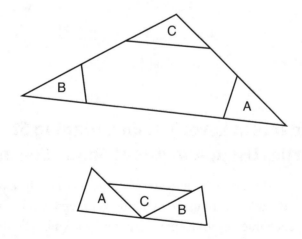

Relating Properties

Once students seem to have a good grasp of the formal properties of shapes (Level 2.3), encourage them to think about how properties are *related*. One of the easiest ways to get students to start relating properties is to have them empirically investigate shapes in dynamic geometry.

Instructional Strategies for Geometric Shapes

For instance, students can use the Measured Quadrilateral Maker to investigate the question, "Can a quadrilateral with opposite sides the same length have opposite sides that are not parallel?" They will find that whenever they make both pairs of opposite sides of the Quadrilateral Maker equal, the opposite sides will be parallel. One way to test for parallelism of the quadrilateral's sides is to construct lines through these sides. Students can then scroll the computer screen to see if the lines meet. Another way is to select both lines and try to use the "construct intersection" command; if the lines are parallel, the command will not be available. Older students can use the computer tool that gives the slopes of lines—two lines are parallel if they have equal slopes. (Intuitively, students can think of slope as the amount of tilt in the line.)

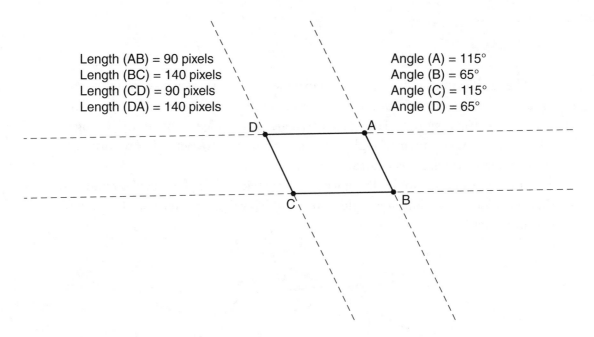

Teaching Students at Level 3.1: Encouraging Students to Relate Properties Using Analysis of Shape Construction

Questions like the two shown below encourage students to analyze relationships between properties. If students are using dynamic geometry to empirically investigate the questions, encourage even deeper thinking by asking why their discoveries are true. Whether they are using dynamic geometry or not, encourage them to investigate the questions by drawing shapes piece by piece.

For instance, consider Question 1.

> **Question 1.** *Is it possible to have a quadrilateral that has 4 right angles, but does not have opposite sides parallel? Write a convincing argument that shows why your answer is correct.*

Suppose a student says, "Whenever I made a 4-sided shape that has 4 right angles (in dynamic geometry or paper-and-pencil drawing), the opposite sides were parallel." Ask the student, "Why do you think that whenever you make a quadrilateral with 4 right angles it has opposite sides parallel? What happens to the angles when the opposite sides are not parallel? Try drawing a quadrilateral that has 4 right angles but does not have opposite sides parallel—what happens?" Such questions can help students move from empirical to more analytic reasoning. Use similar guiding questions to help students on similar tasks.

Question 2. *True or False? If a quadrilateral has opposite sides equal and at least 1 right angle, then the quadrilateral is a rectangle. Write a convincing argument about why your answer is correct.*

Using Construction Tools in Dynamic Geometry Software

One often-used approach in junior high school is to have students use commands to *construct* draggable figures in dynamic geometry software programs. To construct these draggable shapes, students must develop an understanding of how to embed geometric properties in shapes using the software construction tools.

To illustrate how constructing draggable shapes involves students in the analysis of properties and interrelationships between properties, consider the process of constructing a draggable rectangle in *The Geometer's Sketchpad*.

Rectangle Construction 1

Step 1. Use the line segment command to construct segment AB.

Step 2. Construct a line perpendicular to segment AB through point A.

Step 3. Place point C on this perpendicular, and construct segment AC.

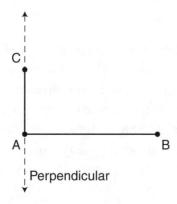

Step 4. Construct a line perpendicular to segment AC through point C.

Step 5. Construct a line perpendicular to segment AB through point B.

Instructional Strategies for Geometric Shapes

Step 6. Construct point D as the intersection of the perpendicular lines from Steps 4 and 5.

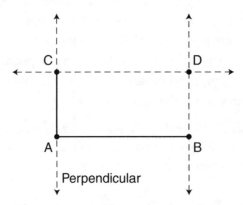

Step 7. Construct segments CD and BD. Hide perpendicular lines.

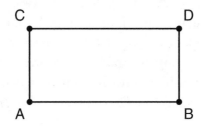

Step 8. Show measurements.

Students might use the measurements created in Step 8 to convince themselves empirically (Levels 2.3 and 3.1 reasoning) that the shape they have constructed always has the properties of a rectangle, even if they drag the vertices. For instance, students can drag the vertices of their shape to verify that the shape always has the properties: "opposite sides equal" and "4 right angles." You can ask questions like, "Are there different ways of constructing a draggable rectangle? What properties are you building into this new construction?" (Rectangle Construction 2 shows one possible way to do this.)

Rectangle Construction 2

Step 1. Use the line segment command to construct segment AB.

Step 2. Construct a line perpendicular to segment AB through point A.

Step 3. Place point C on this perpendicular, and construct segment AC.

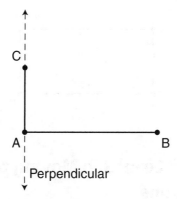

Step 4. Construct the circle with center at point C and radius the length of segment AB.

Step 5. Construct the circle with center at point B and radius the length of segment AC.

Step 6. Make point D the intersection of these two circles.

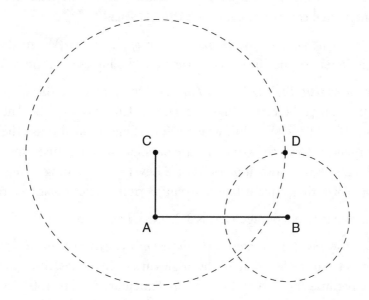

Step 7. Construct segments CD and BD. Hide the circles. Show the measurements.

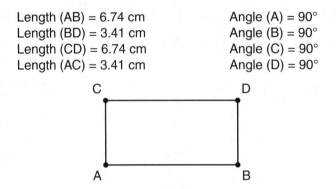

Teaching Students at Level 3.2: Moving to Logical Inference and Minimal Definitions

Logically Analyzing Constructions

One way to encourage students to use Level 3.3 inference is to have them analyze constructions (in dynamic geometry or with paper and pencil) and try to justify or prove that their shape is what they claim. For instance, you can ask the following sequence of questions about the dynamic geometry Rectangle Construction 1 above. (Note that we can ask the same sequence of questions even if we pose the problem with paper-and-pencil constructions—see **STUDENT SHEET 5** .)

What properties did you explicitly construct as part of this shape? [We made Angle A a right angle. Then we specifically constructed right angles at vertices B and C.]

How do we know that Angle D is a right angle? [One way to see this is to use the fact that the sum of the vertex angles in a quadrilateral is 360°. So Angle D measures 360° − 90° − 90° − 90°, which is 90°. (One could also use the fact that 2 lines perpendicular to the same line are parallel, then use theorems on parallel lines and transversals. Both approaches are Level 3.3 reasoning. Note that this reasoning proves that a quadrilateral having 3 right angles must be a rectangle.)

Additional questions to encourage Levels 3.2 and 3.3 reasoning:

Why did the way that you constructed the shape make it have opposite sides parallel? [This comes from the fact that all the angles are right angles. Students might argue that because angles A and C are supplementary (add to 180°), segments AB and CD must be parallel because they learned that when 2 lines cut by a transversal have same side interior angles supplementary, the lines must be parallel. Students might also argue that if 2 lines are perpendicular to the same line, then they must be parallel, which they could have discovered empirically (Level 3.1), reasoned by construction (Level 3.2), or inferred from theorems about parallelism and transversals (Level 3.3).]

Why did the way that you constructed the shape make it have opposite sides equal? [Students might use Level 3.2 reasoning, or they might use the fact that the distance between 2 parallel lines remains constant.]

Are there different ways of constructing a draggable rectangle? What properties are you building into this new construction? [See Rectangle Construction 2 (above) for one possibility. Ask the same questions as described above.]

Additional Tasks to Move Students to Logical Inference

To encourage students to use logical inference, start with straightforward problems that do not require the use of additional theorems, only logical analysis of given statements. For instance, consider these two problems.

> *If a quadrilateral has all sides equal, does it have opposite sides equal? Write a convincing argument that shows why your answer is correct.*
>
> *If a quadrilateral has all sides equal, does it have adjacent sides equal? Write a convincing argument that shows why your answer is correct.*

If need be, you can help students make these inferences with some additional guidance. For instance, for the first question above, you might ask, "Suppose we have a quadrilateral that has all sides equal, can you draw an example? Now, in your example, are the opposite sides equal? Why? Do you think it's possible to draw a shape that has all sides equal but not opposite sides equal? Why?"

As a next step, after students have established empirically and through discussion certain things that everyone in the class agrees must be true, have them infer simple implications from these taken-as-true findings. For instance, after having students establish empirically (especially using dynamic geometry) that 2 lines are parallel if and only if same side interior angles add to 180° (are supplementary), have students do the following two tasks.

> *Is it possible to have a quadrilateral that has 4 right angles, but does not have opposite sides parallel? Write a convincing argument that shows why your answer is correct.*
>
> *True or False? If a quadrilateral has 4 right angles, then the opposite sides of the quadrilateral must be parallel. Write a convincing argument about why your answer is correct.*

You can also give students other kinds of problems that rely on logical analysis.

Instructional Strategies for Geometric Shapes

John claims that if a figure has 4 equal sides, it is a square. Which figure might be used to prove John is wrong? Write a convincing argument about why your answer is correct.

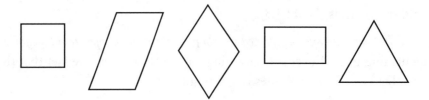

Note that logical inference is required to solve the last problem. The only way to prove that John is wrong is to find a 4-sided figure that is *not* a square. Thus, showing any rhombus that is not a square (such as the middle figure) proves that John is wrong. This figure has 4 equal sides, but it is not a square because it does not have right angles. On this problem, be sure to ask students questions such as "Why does the square on the left not show that John is wrong?" (Some students think that this figure shows that John is right.)

"Why do the second, fourth, and fifth figures not show that John is wrong?"

Justifying Empirical Discoveries

Once students can use logical inference to relate properties, we can encourage them to produce logical justifications for some ideas that they previously investigated empirically or intuitively. For instance, we can have students reconsider the previously discovered finding that the sum of the interior angles of a triangle is 180° with the following task. (In general, Level 3.3 and 3.4 reasoning is an appropriate instructional goal for students in grades 6–8. However, this goal is attainable only if students have interacted with a carefully designed curriculum that has supported their movement to Levels 2.3 and 3.1 in earlier grades.)

> *Assume that we know that 2 lines are parallel if and only if they are cut by a transversal, the alternate interior angles are equal (in measure). Use this fact to prove that the sum of the interior angles of a triangle is 180°. The diagram below will help you. Line L is parallel to Line AB.*

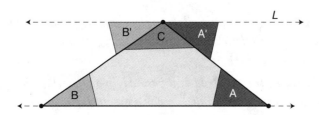

Expected student justification: Because lines AB and L are parallel, angles B and B′ are equal, and angles A and A′ are equal. But angles B′, C, and A′ form a straight angle, so their measures sum to 180°. So the sum of angles A, B, and C is 180°.

Another problem that students can investigate is showing that the sum of the interior angles of a polygon with n sides is (n–2)(180°). Students can justify this theorem by decomposing polygons into an appropriate set of n–2 triangles.

Moving Students to Minimal Definitions

Once students seem capable of making simple inferences to relate shape properties, encourage them to think about minimal definitions for types of shapes and hierarchical classification.

> *(a) Describe all the properties that a rectangle must have.*
>
> *(b) What's the smallest set of properties that you can name to be sure that a quadrilateral is a rectangle? What one property of rectangles can you choose that will force another property to occur? How do you know that a quadrilateral with this set of properties will have all the properties you listed in part (a)?*

To help students, recognize the kind of reasoning they can build on. For instance, these minimal definition tasks are appropriate for students who are capable of Level 3.1, 3.2, and 3.3 reasoning. These students can analyze these questions empirically (especially using dynamic geometry), through constructive analysis in drawing shapes, and through logical implication. For instance, in thinking about a minimal definition for a rectangle, students might claim, "If a quadrilateral has 4 right angles, then I know it has opposite sides parallel because of what we learned about parallels and the lines that intersect them *[transversals]*. And if it has 4 right angles and the opposite sides parallel, it's opposite sides must be equal."

> **Teacher:** *How do you know the opposite sides will be equal?*
>
> **Student:** *Well, if you try to draw a quadrilateral with 4 right angles and opposite sides parallel, you just can't make the opposite sides unequal* [Level 3.2].
>
> **Teacher:** *OK, good. Are there other ways to think about this? What do you call a quadrilateral that has opposite sides parallel?*
>
> **Student:** *A parallelogram.*
>
> **Teacher:** *Right. Now what happens in a Measured Parallelogram Maker on the computer if we make the angles 90°, what kind of shape is it? Explore this on the computer and see what you find.*

Another way to help this student if she has learned some of the construction tools in dynamic geometry is to ask her to construct a draggable quadrilateral that has opposite sides parallel and 90° angles. Then ask, "What happens to the opposite side lengths in this construction?"

Instructional Strategies for Geometric Shapes

(a) Describe all the properties that a parallelogram must have.

(b) What's the smallest set of properties that you can name to be sure that a quadrilateral is a parallelogram? What one property of parallelograms can you choose that will force another property to occur? How do you know that a quadrilateral with this set of properties will have all the properties you listed in part (a)?

(a) Describe all the properties that a kite must have.

(b) What's the smallest set of properties that you can name to be sure that a quadrilateral is a kite? What one property of kites can you choose that will force another property to occur? How do you know that a quadrilateral with this smallest set of properties will have all the properties you listed in part (a)?

Does the following definition give enough information to guarantee that a shape is a square? "A square is a quadrilateral with 4 congruent sides and at least 1 right angle."

Give a definition for a rectangle that does not list all of its properties.

Teaching Students at Level 3.3: Moving to Hierarchical Classification

Having whole-class discussions about the set of True/False questions below always seems to help students reason productively about classification questions. (Note, however, that although students will make progress in their reasoning during such discussions, some students will still not completely accept the conclusions about classification questions, especially about squares and rectangles.) Have separate class discussions about successive pairs of problems. Ask students to give an answer (true/false), then try to convince their classmates that they are correct.

As students argue, guide the discussions to consideration of properties. For instance, for Pair 3 you can ask, "What are all the properties of squares? Do all rectangles have all these properties? What are all the properties of rectangles? Do all squares have these properties?"

True or False, and Why?

Pair 1. All rhombuses are parallelograms.
All parallelograms are rhombuses.

Pair 2. All kites are rhombuses?
All rhombuses are kites?

Pair 3. All squares are rectangles.
All rectangles are squares.

One way to help students understand these ideas is to have them use dynamic geometry constructions that show their measurements. For instance, students can be asked to investigate Pair 3 above using a Measured Rectangle Maker and a Measured Square Maker.

Rectangle Maker

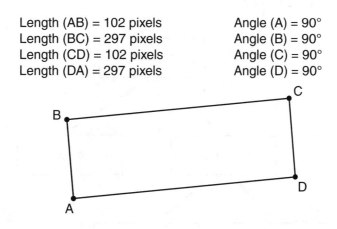

Length (AB) = 102 pixels   Angle (A) = 90°
Length (BC) = 297 pixels   Angle (B) = 90°
Length (CD) = 102 pixels   Angle (C) = 90°
Length (DA) = 297 pixels   Angle (D) = 90°

Square Maker

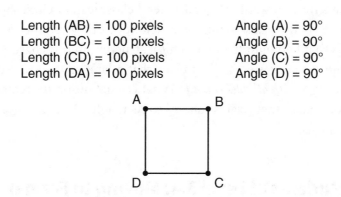

Length (AB) = 100 pixels   Angle (A) = 90°
Length (BC) = 100 pixels   Angle (B) = 90°
Length (CD) = 100 pixels   Angle (C) = 90°
Length (DA) = 100 pixels   Angle (D) = 90°

After students have convinced themselves that the Rectangle Maker will make all squares but the Square Maker will not make all rectangles, have students use logical analysis to understand the situation. For instance, ask students, "What are all the properties of squares, and of rectangles? What's the definition for a rectangle? Do squares satisfy this definition? Are squares rectangles?"

Have students compare the following two definitions for rectangles.

Definition 1: A rectangle is a quadrilateral with opposite sides equal and 4 right angles.

Definition 2: A rectangle is a quadrilateral with opposite sides equal and 4 right angles, but not all sides equal.

Ask questions such as "Which definition do you think is better? Why? For which definition are squares rectangles? Suppose that I figure out that in rectangles the diagonals bisect each other; for which definition would my finding also apply to squares?"

Create a classification chart for quadrilaterals that shows relationships between different kinds of quadrilaterals. For instance, if we were thinking about animals, part of our chart might look like this.

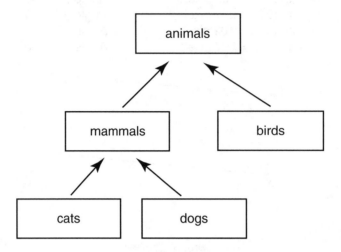

Help students move toward the quadrilateral classification chart shown in Chapter 2 (page 44). Also ask students what definitions they can use for classes of quadrilaterals that highlight the hierarchical structure in this chart. "What is a definition for squares suggested by your chart that relates squares to rectangles?" *[A square is a rectangle with all sides equal.]* "What is a definition for rectangles that is suggested by your chart, that relates squares to rectangles?" *[A rectangle is a parallelogram with right angles.]*

Teaching Students at Level 3.4: Moving to Formal Deductive Proofs

Extensive discussion of how to move students to formal proof is beyond the grade-level scope of the CBA materials. See, for example, the *Discovering Geometry* textbook by Serra (2003) or the UCSMP high school geometry textbook (Usiskin, Hirschorn, & Coxford, 2003) for extended treatments.

However, here is one example of a transitional activity you can use with students at grades 7 and 8. In this activity, students assume certain statements are true, then try to prove another statement is true.

Assume the following theorem is true.

Theorem: *If 2 lines (m and n) are intersected by a line (k) so that the same side interior angles (F and C) are supplementary (add to 180°), then the 2 lines (m and n) are parallel.*

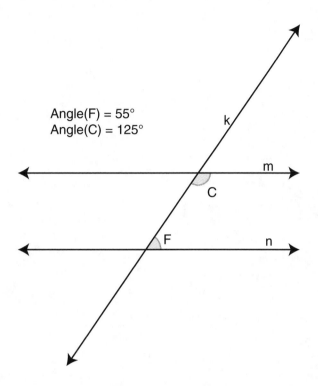

Use this theorem to prove that: If a quadrilateral has 4 right angles, then the opposite sides of the quadrilateral must be parallel.

In essence, most activities that promote reasoning at Levels 3.3 and 3.4 are excellent for preparing students for high school geometry. Students who are *fluent* in Level 3.3 and 3.4 reasoning have the intellectual tools that can enable them to make sense of and succeed in a formal treatment of geometry.

Appendix

CBA Assessment Tasks for Geometric Shapes

These problems can be used in individual interviews with children or in class as instructional activities. However, no matter which approach you choose, it is critical to get the students to describe and discuss their strategies. Only then can you use the CBA levels to interpret students' responses and decide on needed instruction.

Guide for Interviewing Students with CBA Tasks

The purpose of interviewing students with CBA tasks is to determine how they are reasoning and, more specifically, to determine what CBA levels of reasoning students are using for the tasks.

Before the interview, CBA teachers said the following to students:

> I am going to give you some problems. I would like to know what you think while you solve these problems. So, tell me everything you think as you do the problems. Try to think out loud. Tell me what you are doing and why you are doing it. I will also ask you questions to help me understand what you are thinking. For instance, if you say something that I don't understand, I will ask you questions about it.

If you don't understand what a student is saying, you could ask, "I don't understand, could you explain that again?" or "What do you mean by such-and-such?" Try to get students to explain in their own words, rather than paraphrasing what you think they mean and asking if they agree. If, during an interview, a student asks whether his or her answers are correct, we told the student that, for this interview, that does not really matter. We are interested in what he or she thinks.

Students responded to our request to "think out loud" in two ways. Many students were quite capable of thinking out loud as they solved problems. They told us what they were thinking and doing as they thought and did it. Other students, however, seemed unable to think aloud as they completed problems. They worked in silence, but then gave us detailed accounts of what they did *after* they finished doing it.

The following tasks cover a large range of geometric reasoning. You probably will not want to give all the problems to your students, at least not at one time. For students in grades 1–3, it is suggested that you select from Problems 1–5. For students in grades 4–8, it is suggested that you select from Problems 1–9. Of course, you can alter these suggestions based on your curriculum.

Problems 1–7 are especially helpful in deciding whether students' reasoning is in CBA Level 1 or 2. These problems can also help you decide which sublevel of reasoning students are using within Levels 1 and 2. Problems 8 and 9 are especially helpful for deciding sublevels within Level 3. If students ignore the measurements in Problems 6 and 7, there is no need to give them Problems 8 and 9.

Many of the problems have notes that indicate particular aspects of students' reasoning emphasized by the problems.

Additional assessment tasks may be downloaded from this book's website, www.heinemann.com/products/E04351.aspx. (Click on the "Companion Resources" tab.)

Name _____   Date _____

1. Drawing Machine 1 made these shapes. All shapes that can be made by Drawing Machine 1 are alike in some way. They follow a rule.

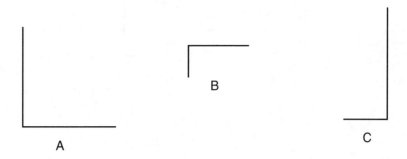

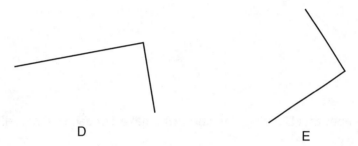

a. Describe how the shapes made by Drawing Machine 1 are alike. Tell what the rule is for this drawing machine.

May be photocopied for classroom use. © 2012 by Michael Battista from *Cognition-Based Assessment and Teaching of Geometric Shapes: Building on Students' Reasoning*. Portsmouth, NH: Heinemann.

Appendix

Name _____ Date _____

b. Drawing Machine 1 *cannot* make this shape. Why not?

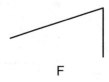

F

c. Circle the shapes that Drawing Machine 1 *can* make. For each shape, tell why Drawing Machine 1 can or cannot make it.

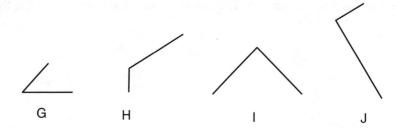

G　　　　H　　　　　I　　　　　J

Name _____ Date _____

2. Drawing Machine 2 made these shapes. All shapes that can be made by Drawing Machine 2 are alike in some way. They follow a rule.

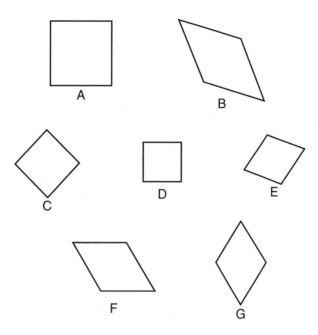

a. Describe how the shapes made by Drawing Machine 2 are alike. Tell what the rule is for this drawing machine.

Name _____ Date _____

b. Drawing Machine 2 *cannot* make this shape. Why not?

H

c. Circle the shapes that Drawing Machine 2 *can* make. For each shape, tell why Drawing Machine 2 can or cannot make it.

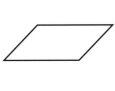 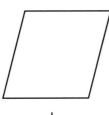   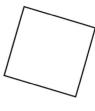

I J K L

Name _____ Date _____

3. Drawing Machine 3 made these shapes. All shapes that can be made by Drawing Machine 3 are alike in some way. They follow a rule.

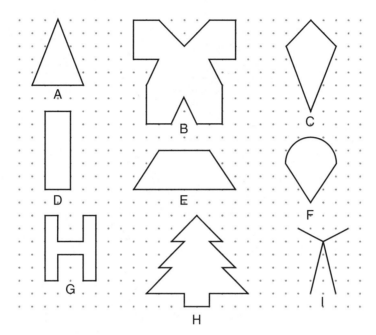

a. Describe how the shapes made by Drawing Machine 3 are alike. Tell what the rule is for this drawing machine.

May be photocopied for classroom use. © 2012 by Michael Battista from *Cognition-Based Assessment and Teaching of Geometric Shapes: Building on Students' Reasoning*. Portsmouth, NH: Heinemann.

Appendix 117

b. Drawing Machine 3 *cannot* make these shapes. Why not?

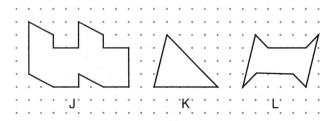

c. Circle the shapes that Drawing Machine 3 *can* make. For each shape, tell why Drawing Machine 3 can or cannot make it.

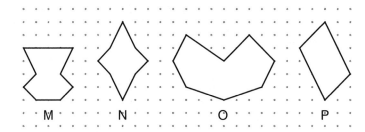

Name _____   Date _____

4. a. Circle each triangle.

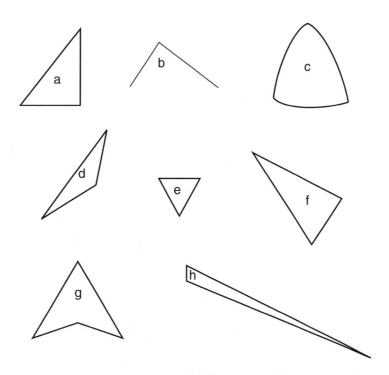

b. Describe exactly how you decide if a shape is a triangle or not.

May be photocopied for classroom use. © 2012 by Michael Battista from *Cognition-Based Assessment and Teaching of Geometric Shapes: Building on Students' Reasoning*. Portsmouth, NH: Heinemann.

c. Is Shape *d* a triangle? Explain why.

d. Describe everything you know about triangles.

Name _____ Date _____

5. a. Circle each rectangle.

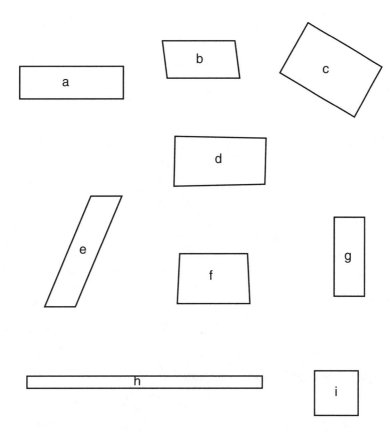

b. Describe exactly how you decide if a shape is a rectangle or not.

c. Is Shape *c* a rectangle? Explain why.

d. Describe everything you know about rectangles.

Name _____   Date _____

6. The measurements for a shape are given below.

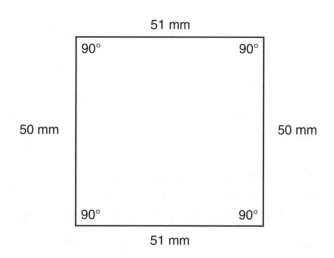

For each statement about this shape, circle *T* if the statement is True, or *F* if the statement is False.
If you can't tell if the statement is true or false, circle *Can't tell*.
For each statement, describe how you would convince someone that your answer is correct.

    a. The shape is a square.   T   F   Can't tell

    b. The shape is a rectangle.   T   F   Can't tell

    c. The shape is a parallelogram.   T   F   Can't tell

    d. The shape is a rhombus.   T   F   Can't tell

May be photocopied for classroom use. © 2012 by Michael Battista from *Cognition-Based Assessment and Teaching of Geometric Shapes: Building on Students' Reasoning*. Portsmouth, NH: Heinemann.

Name _____   Date _____

7. The measurements for a shape are given below.

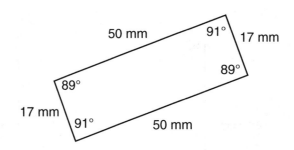

For each statement about this shape, circle *T* if the statement is True, or *F* if the statement is False.
If you can't tell if the statement is true or false, circle *Can't tell*.
For each statement, describe how you would convince someone that your answer is correct.

    a. The shape is a square.   T   F   Can't tell

    b. The shape is a rectangle.   T   F   Can't tell

    c. The shape is a parallelogram.   T   F   Can't tell

May be photocopied for classroom use. © 2012 by Michael Battista from *Cognition-Based Assessment and Teaching of Geometric Shapes: Building on Students' Reasoning*. Portsmouth, NH: Heinemann.

Name _____ Date _____

8. A quadrilateral is a closed shape with four straight sides. Examples of quadrilaterals are squares, rectangles, rhombuses, parallelograms, kites, and trapezoids.

Circle *True* if the statement is True, or *False* if the statement is False.
Describe how you would prove or show that your answer is correct.

If a quadrilateral has opposite sides equal and at least one right angle, then the quadrilateral is a rectangle.

Circle One: True   False   *Prove your answer or tell why your answer is correct.*

Name _____ Date _____

9. Tell whether the statement in the box is true or false.

Circle your answer.   True   False

All squares are rectangles.

What would you say to convince other students that your answer is correct?

CBA Levels for Each Task

The descriptions below show sample student responses at various levels of sophistication.

PROBLEM 1

Drawing Machine 1 made these shapes. All shapes that can be made by Drawing Machine 1 are alike in some way. They follow a rule.

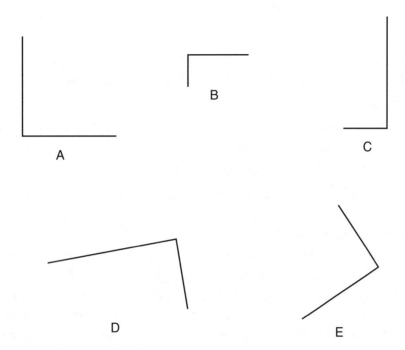

(a) Describe how the shapes made by Drawing Machine 1 are alike. Tell what the rule is for this drawing machine.

(b) Drawing Machine 1 *cannot* make this shape. Why not?

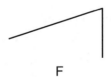

CBA Assessment Tasks for Geometric Shapes

(c) Circle the shapes that Drawing Machine 1 *can* make. For each shape, tell why Drawing Machine 1 can or cannot make it.

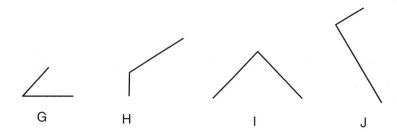

Level 1.1: (a) I think what is alike about them is that they all kind of bend. (b) I don't know. (c) *[Circles Shape G]* I think this one can be made by the drawing machine.

Level 1.2: (a) They are all shaped like L's. (b) It's not an L. (c) *[Loops Angles I and J]* Because if you look at it *[I]* this way *[rotates paper]*, it looks like an L, and if you look at these two *[H and G]*, they're diagonal.

Level 2.1: (a) They all got 1 point *[pointing at the vertices of the angles]*. They all have 2 strings, well, lines. They all have 1 point and they all have 2 lines. (b) *[Turns the student sheet 90°, then 90° again]*. I am thinking. It has 2 lines like the other shapes *[A–E]*. But I can't think why it can't be made by the drawing machine. Or: Because Shape F has a diagonal line *[uses his pen to trace over the top ray of angle F]*, and none of those *[Shapes A–E]* have a diagonal line. They're just in a different angle *[motions a turn with his hand from the wrist]*. (c) *[Circles only Shape I]* Because it doesn't have a slant. Because if you turn it *[turning the paper 90°]*, it doesn't have a slant. If you turn this *[Shape H]*, it has a slant, and that one *[Shape G]*. This one has a slant *[crossing out Shape G]*. This one has a slant *[crossing out Shape H]*. And this one doesn't have a slant *[Shape I]*. (*Note:* by "slant," the student appears to mean perpendicular, which is the common attribute of Shapes A–E, but the word "slant" is informal and imprecise.)

Level 2.2: (a) They all look like right angles. (b) Because the rule for this one is always straight and straight, and this one's slanted. So this one can't be because it's an acute angle. *[Teacher: And the others were?]* Right. (c) *[Circles I and J]* This one *[G]* is an acute angle, it's too small, and this one's *[H]* too big, it's an obtuse angle, but this one *[I]* if you look at it like this *[turns student sheet so that one ray is parallel to edge of table]* is straight and straight. *[Teacher: And that one (I) is?]* A right angle.

The student correctly uses the formal concepts of right, acute, and obtuse angle to reason about the problems. This example is included in this section because it suggests how students sometimes make personal sense of the formal concepts of right and acute angles. The student's informal description "straight and straight, and this one's slanted" to distinguish right angles from acute angles suggests that his formal ideas may be rooted in his earlier informal ideas.

Level 2.3: (a) They are all 90° angles. (b) Because it is not a 90° angle. (c) *[Circles I and J]* Cause if you turn the paper *[rotates student sheet]*, you can see that it *[I]* is a 90° angle.

The student's identification of shapes is completely based on the use of a formal geometric property (having right angles).

Levels 3.1–4: Not applicable.

PROBLEM 2

Drawing Machine 2 made these shapes. All shapes that can be made by Drawing Machine 2 are alike in some way. They follow a rule.

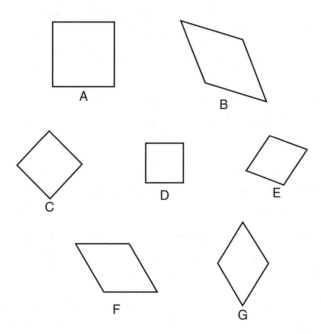

(a) Describe how the shapes made by Drawing Machine 2 are alike. Tell what the rule is for this drawing machine.
(b) Drawing Machine 2 *cannot* make this shape. Why not?

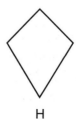

CBA Assessment Tasks for Geometric Shapes

(c) Circle the shapes that Drawing Machine 2 *can* make. For each shape, tell why Drawing Machine 2 can or cannot make it.

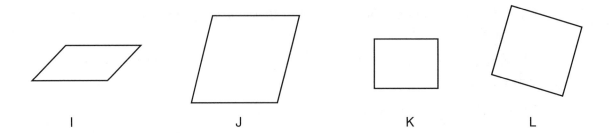

Level 1.1: (a) There are 3 different shapes here. *[Shapes A, D, and E]* It's *[pointing at A]* the same as this *[pointing at D]*. Shape E is like Shape D because you could just flip it *[E]* over like this, so they would be the same. *[Shapes C and G]* This *[pointing at C]* is exactly like this *[pointing at G]* only it's wider. This *[G]* is longer. *[Shapes B and F]* These are the same but one *[F]* is bigger and wider. *[Teacher: So what's the rule for these shapes?]* The rule is that it makes them larger and fatter. (b) I don't know. (c) I and J maybe.

Typical of students at this level, this student was unable to see the common characteristic of all the shapes shown above (that each one is a rhombus and thus has all 4 sides the same length).

Level 1.2: (a) They all kind of look like diamonds. *[Teacher: How do you know those are diamonds?]* Because if you turned them they would look like diamonds. (b) It looks like a kite, but no other shape looks like a kite. (c) Probably J and L.

Even though "diamond" is an imprecise, holistic description of the example shapes, this student's conceptualization is sufficient to correctly identify examples and non-examples of shapes like A–G.

Level 2.1: (a) They all have 4 lines or a slant. Not a slant. Four points *[pointing to each of the four vertices for a few of Shapes A–G]*. *[Teacher: So, what do you think the rule is?]* They all have 4 lines and 4 slants. I mean not 4 slants, 4 points. (b) I don't know. (c) Probably I and J [or J and L].

This student's informal parts-based description is not only imprecise, it does not deal at all with the defining characteristic of Shapes A–G—that all sides are equal in length (so the shapes are all rhombuses).

Level 2.2: (a) The shape has to be symmetrical, the sides across have to be parallel, and it also has to be equal on all sides, and it also has to have 4 sides. (b) H doesn't have all sides equal. (c) *[Chooses J, K, and L]* These two *[K and L]* have 4 equal sides, and are parallel, and are symmetrical. And this one *[J]* has 4 equal sides and is symmetrical.

Although all the properties the student described for Shapes A–G are formal and correct, he said that Shape K has 4 equal sides, which is incorrect. So his incorrect identification of shapes was inconsistent with his correct formal description.

Level 2.3: (a) The shape has to have 4 equal sides. (b) H doesn't have all sides equal. (c) *[Chooses J and L]* Change L to make it clear that it is a square.

Levels 3.1–3.2: Not applicable.

Level 3.3: (a) They all have at least 2 lines of symmetry. The squares go up like this and this *[draws a horizontal and vertical line of symmetry on Shape A]* and these *[draws diagonals on F]* go like this and like this. But the squares can have more than 2 lines of symmetry, so they *[Shapes A–G]* should have *at least* 2 lines of symmetry because this shape *[F]* can only have 2 lines of symmetry. (b) It does not have 2 lines of symmetry. (c) J and L have at least 2 lines of symmetry.

This student made one of the simplest kinds of logical connections of properties; he connected the properties "have 2 lines of symmetry," and "have more than 2 lines of symmetry" with the statement, "have at least 2 lines of symmetry."

Levels 3.4–4: Not applicable.

PROBLEM 3

Drawing Machine 3 made these shapes. All shapes that can be made by Drawing Machine 3 are alike in some way. They follow a rule.

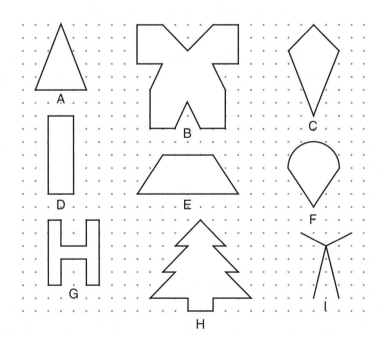

(a) Describe how the shapes made by Drawing Machine 3 are alike. Tell what the rule is for this drawing machine.

CBA Assessment Tasks for Geometric Shapes

(b) Drawing Machine 3 *cannot* make these shapes. Why not?

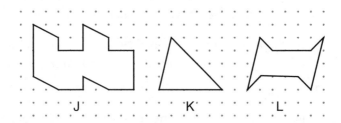

(c) Circle the shapes that Drawing Machine 3 *can* make. For each shape, tell why Drawing Machine 3 can or cannot make it.

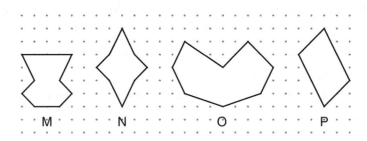

Level 1.1: (a) They're all standing up. (b) J and K look like they're standing up, but L is tilted. (c) All of them. They're all standing up.

Level 1.2: (a) They look sort of like perfect shapes. They're even. (b) They're not even. (c) I'd say M and N can be drawn, but O and P can't because they're not perfect.

Level 2.1: (a) Both sides of the shapes are the same, see, like the right side and the left side. (b) The right and left sides are different. (c) M and N look like the sides are the same. So does P.

Level 2.2: (a) You can fold them so the right and left sides go onto each other. (b) You can't fold them so the sides match. (c) M and N, you can fold. But O and P you can't.

Level 2.3: (a) They all have vertical lines of symmetry. The left side flips onto the right side and matches exactly. (b) They don't have vertical lines of symmetry. (c) M and N have vertical lines of symmetry, but O and P don't.

Levels 3.1–3.2: Not applicable.

Level 3.3: (a–c) They all have vertical lines of symmetry. The left side reflects onto the right side and matches exactly. And if you make a line segment join a point on one side of the symmetry line to its reflection on the other side, the symmetry line is the perpendicular bisector of the segment.

Level 3.4: Not applicable.

Level 4: Would require a proof of the statement made in Level 3.3.

PROBLEM 4

(a) Circle each triangle.

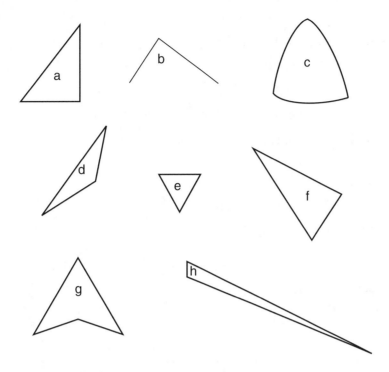

(b) Describe exactly how you decide if a shape is a triangle or not.
(c) Is Shape d a triangle? YES   NO   Explain why.
(d) Describe everything you know about triangles.

Levels 1.1–1.2: (a) Students have to make triangle/not-triangle decisions for 8 shapes. Students who correctly decide for at least 7 shapes can be classified as Level 1.2 if their reasoning is holistic-visual as described below. (b, c, and d)

For parts (b), (c), and (d), students will make statements like the following:

It looks like (doesn't look like) a triangle.

It's pointy.

Because if you turn the shape, it will look like a triangle.

It looks like Shape a.

[For Shape h] It just doesn't look like a triangle. It is just like a point.

Level 2.1: (a) *[Circles a, c, d, e, f.]* (b) They have 3 points. (c) Because it has 3 points, same as *[Shapes]* d, e, and f. *[May not identify h or d as a triangle.]* (d) They have 3 points.

CBA Assessment Tasks for Geometric Shapes

The student uses the formal term "points" incorrectly to refer to vertices or the "pointy parts" of triangles. Formally, a point is a location in space. Triangles have an infinite number of points—every location on the sides of a triangle is a point on the triangle.

Level 2.2: (a) *[Circles a, d, e, f]* (b) They have 3 points and 3 sides. (c) Yes, because it has 3 points and sides. (d) They have 3 points and 3 sides.

Level 2.3: (a) *[Circles a, d, e, f, h]* (b) They have 3 points. (c) Yes because it has 3 straight sides. (d) Triangles have 3 straight sides, and they are closed.

Levels 3.1–4: Not applicable.

PROBLEM 5

(a) Circle each rectangle.

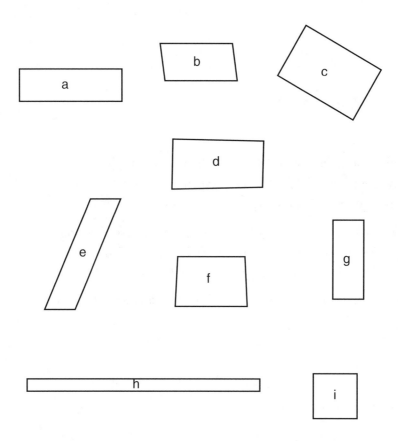

(b) Describe exactly how you decide if a shape is a rectangle or not.
(c) Is Shape *c* a rectangle? YES NO Explain why.
(d) Describe everything you know about rectangles.

Level 1.1: (a) *[Circles a, b, d, e, g, and h]* (b) They're sort of straight. (c) No, it's slanted. (d) Rectangles stand up straight or they lay down. Or: a square is longer than a rectangle.

Level 1.2: (a) Student correctly says that Shapes a, c, g, and h are rectangles, and does not correctly decide that Shape i (a square) is a rectangle. May include one or two nonrectangles. (b) They're longer than squares. (c) Yes, it's longer than a square. (d) Rectangles look like a door or a table.

Level 2.1: (a) Student correctly says that Shapes a, c, g, and h are rectangles, but also includes perhaps b, e, or f. (b) They have 2 long sides and 2 short sides. (c) Yes, it has 2 long sides and 2 short sides. (d) Rectangles have 2 long sides and 2 short sides.

Level 2.2: (a) Student correctly says that Shapes a, c, g, and h are rectangles but does not correctly decide that Shape i (a square) is a rectangle. (b) The sides across from each other are equal, and it has square corners. (c) Yes, the sides across from each other are equal, and it has square corners. (d) In a rectangle, the sides across from each other are equal, and they have square corners.

Level 2.3: (a) Student correctly says that Shapes a, c, g, and h are rectangles but does not correctly decide that Shape i (a square) is a rectangle. (b) Opposite sides equal and 4 right angles. (c) Yes, it has opposite sides equal and 4 right angles. (d) To be a rectangle, opposite sides must be equal, and it has to have 4 right [90°] angles.

Levels 3.1–3.2: Not applicable.

Level 3.3: (a) Student correctly says that Shapes a, c, g, and h are rectangles. *[May not correctly decide that Shape i, which is a square, is a rectangle.]* (b) Opposite sides equal and 4 right angles. (c) Yes, it has opposite sides equal and 4 right angles. (d) In a rectangle, the sides across from each other are equal, and they have 4 right [90°] angles. But actually you could say a rectangle has 4 sides and 4 right angles. Having 4 right angles means the opposite sides will be congruent too.

Level 3.4: (a) Student correctly says that Shapes a, c, g, h, and i are rectangles. (b) Opposite sides equal and 4 right angles. (c) Yes, it has opposite sides equal and 4 right angles. (d) Rectangles are parallelograms with right angles. Shape i is a rectangle because squares are rectangles with equal sides.

Level 4: Not applicable.

PROBLEM 6

The measurements for a shape are given below.

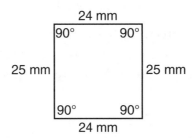

For each statement about this shape, circle *T* if the statement is True, or *F* if the statement is False.

If you can't tell if the statement is true or false, circle *Can't tell*.

For each statement, describe how you would convince someone that your answer is correct.

(a) The shape is a square.   T   F   Can't tell
(b) The shape is a rectangle.   T   F   Can't tell
(c) The shape is a parallelogram.   T   F   Can't tell
(d) The shape is a rhombus.   T   F   Can't tell

Levels 1.1–1.2: (a) True. Yes, it looks like a square. (b) False. No, because it is not long. (c, d) I don't know.

Level 2.1: (a) True. Yes, it's a square. The sides look the same. (b) False. No, because it does not have 2 long sides and 2 short sides. (c, d) I don't know.

Level 2.2: (a) False. No, I don't think it is a square because a square has all equal sides and this is 24 and this is 25. So I think it would be false. (b) False. No, because it does not have 2 long sides and 2 short sides. (c) No, it does not have 2 long sides and 2 shorts sides, and slants. (d) No, it looks like it has all sides equal, but it doesn't have slanty sides.

Level 2.3: (a) False. No, because a square has all equal sides and this is 24 and this is 25. So I think it would be false. (b) True. Yes, because opposite sides are equal (congruent) and it has 4 right angles. (c) I don't know. (d) False. I think it's not a rhombus because all rhombuses have all equal sides.

Levels 3.1–3.2: Not applicable.

Level 3.3: (a) False. No, because a square has all equal sides and this is 24 and this is 25. So I think it would be false. (b) True. Yes, because opposite sides are equal (congruent) and it has 4 right angles. (c) True. Yes, because opposite sides are equal (congruent) and parallel. (d) False. No, because rhombuses have all equal sides.

Level 3.4: (a) False. No, because a square has all equal sides and this is 24 and this is 25. So I think it would be false. (b) True. Yes, because opposite sides are equal (congruent) and it has 4 right angles. (c) True. Yes, because it's a rectangle and rectangles are parallelograms. (d) False. No, because rhombuses have all equal sides.

Level 4: Not applicable.

PROBLEM 7

The measurements for a shape are given below.

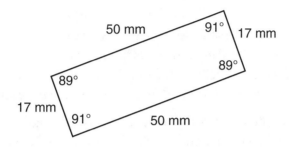

For each statement about this shape, circle *T* if the statement is True, or *F* if the statement is False.

If you can't tell if the statement is true or false, circle *Can't tell*.

For each statement, describe how you would convince someone that your answer is correct.

(a) The shape is a square.   T   F   Can't tell
(b) The shape is a rectangle.   T   F   Can't tell
(c) The shape is a parallelogram.   T   F   Can't tell

Levels 1.1–1.2: (a) False. No, it's too long to be a square. (b) True. Yes, it looks like a rectangle. (c) Don't know what a parallelogram is.

Level 2.1: (a) False. No, these sides are too long to be a square. (b) True. Yes, it has 2 long sides and 2 short sides. (c) Don't know what a parallelogram is.

Level 2.2: (a) False. No, the sides are not equal, and they have to be in a square. (b) True. Yes, because the sides across from each other are equal. (c) The sides look like they are parallel because I don't think that the lines through them would ever intersect.

Level 2.3: (a) False. No, the sides are not equal, and they have to be in a square. (b) False. No, because this shape doesn't have four 90° angles. (c) True. Yes, because opposite sides are equal and parallel.

Level 3.1: (a) False. No, the sides are not equal, and they have to be in a square. (b) False. No, because this shape doesn't have four 90° angles. (c) True. It's a parallelogram because we found out in our explorations on the computer that if a quadrilateral has opposite sides equal its opposite sides are parallel.

Level 3.2: (a) False. No, the sides are not equal, and they have to be in a square. (b) False. No, because a rectangle has to have 90°; this shape doesn't have all 90° angles, here and here and here and here *[pointing to each vertex]* it has 91, 91, 89, and 89. *[Teacher: Why do you think that rectangles have all 90 degree angles?]* Because they wouldn't be rectangles if they didn't have all 90° angles? *[Teacher: Why wouldn't they be rectangles if they didn't have all 90° angles?]* Because if the angles were too far *[points to the vertices of the figure as he is explaining]*, it would be a parallelogram. If they were too far inward or too far out here, it wouldn't be a rectangle, it would be a parallelogram. (c) Yes. Because the opposite sides are equal. If you draw a quadrilateral that has both pairs of opposite sides equal, then the opposite sides are parallel too. If 2 of the sides are not parallel, it makes the other pair of sides not equal *[student illustrates by making a drawing step-by-step]*.

The student not only specifies that rectangles have all angles of 90°, he explains why. In saying that if the angles were not 90° they would be "too far inward or too far out," the student makes explicit how his intuitive understanding of the spatial relationship between adjacent sides in a rectangle is connected to the formal concept of 90° angles. His explanation is framed in terms of a step-by-step construction of the shape.

Level 3.3: (a) False. No, the sides are not equal, and they have to be in a square. (b) False. No, because a rectangle has to have 90°; this shape doesn't have all 90° angles, here and here and here and here *[pointing to each vertex]* it has 91, 91, 89, and 89. (c) True. It's a parallelogram because each pair of consecutive angles is supplementary, which makes opposite sides parallel. We had a theorem that told us that if the same side interior angles formed by a transversal were supplementary, the lines are parallel.

Levels 3.4–4: Not applicable.

PROBLEM 8

A quadrilateral is a closed shape with 4 straight sides. Examples of quadrilaterals are squares, rectangles, rhombuses, parallelograms, kites, and trapezoids.

For the statement below, circle *True* if the statement is True, or *False* if the statement is False.

Describe how you would prove or show that your answer is correct.

If a quadrilateral has opposite sides equal and at least one right angle, then the quadrilateral is a rectangle.

Circle One: True   False   *Prove your answer or tell why your answer is correct.*

Levels 1.1–2.1: Not applicable.

Level 2.2: I don't know.

Level 2.3: No, because rectangles have to have 4 right angles.

Level 3.1: True. All the shapes that I can think of with opposite sides equal, when they have a right angle, they have all right angles.

Level 3.2: It's true. Because, say, there's a 90° angle right here *[draws a 90° angle; see Figure 1]*

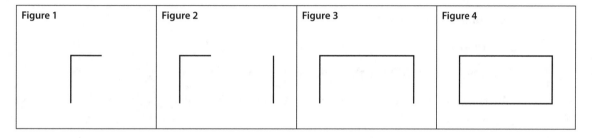

This side right here *[left side]* has to be equal to this side right there *[draws a vertical line segment opposite the left side that is already drawn; see Figure 2]*. And it has to be lined up with the top of the right angle, like across here *[extends the partial segment across the top to meet the right side; see Figure 3]*. If it's not lined up, it won't be a right angle here *[pointing to the upper right angle]*.

And you get right angles on the bottom because this side across the bottom has to be equal to the top side *[draws the bottom side; see Figure 4]*. So it fits in to make right angles; if it's not the same as the top, then you don't get right angles.

So you get a rectangle, which has four 90° angles.

Based on her analysis of the *process* of constructing a quadrilateral with the given properties, part-by-part, the student argued that if you construct a quadrilateral having 1 right angle and opposite sides equal you *must* get a rectangle, which has four 90° angles. (Note that there are logical gaps in her argument.)

Level 3.3: We learned that if a quadrilateral has opposite sides equal it has opposite sides parallel. We also learned that if lines are parallel, then the inside interior angles add up to 180. So, if you have 1 right angle, then because the sides are parallel, the adjacent angles must be 90, which will make the fourth angle 90 also.

Level 3.4: We learned that if a quadrilateral has opposite sides equal it has opposite sides parallel. So this figure is a parallelogram. But a rectangle is a parallelogram with at least 1 right angle (but then it has all right angles).

Level 4: Proof:

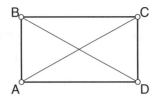

We are given that opposite sides of quadrilateral ABCD are congruent (equal in length). This means that:

AB is congruent to DC.

BC is congruent to AD.

BD is congruent to BD (by the reflexive property of congruence).

So, by the Side-Side-Side congruence theorem, triangle BAD is congruent to triangle BCD.

We are also given that 1 angle is a right angle; suppose this angle is angle BAD.

Because corresponding parts of congruent triangles are congruent, angle BCD must also be a right angle. So we now know that angles BAD and BCD are both right angles, and thus sum to 180°.

Using a similar argument, we get that triangle ABC is congruent to triangle ADC, and that angle ABC is congruent to angle ADC.

Because the sum of the angles in a quadrilateral is 360°, and the measures of angles BAD and BCD sum to 180°, we know that angles ABC and ADC sum to 180°, and are thus both right angles. So all the angles of quadrilateral ABCD are right angles.

Because angle ABC plus angle BAD equals 180°, side BC is parallel to side AD. This is because when lines BC and AD are cut by the transversal AB, the same side interior angles, angle ABC and angle BAD, are supplementary, which implies that BC is parallel to AD.

Using similar reasoning, because angle ABC plus angle BCD equals 180°, side AB is parallel to side CD.

So the opposite sides of quadrilateral ABCD are parallel.

PROBLEM 9

Tell whether the statement in the box is true or false.

Circle your answer.   True   False

> All squares are rectangles.

What would you say to convince other students that your answer is correct?

Levels 1.1–1.2: False. Squares are squares and rectangles are like long squares.

Level 2.1: False. Rectangles have 2 long sides and 2 short sides. All the sides in a square are even.

Level 2.2: False. Rectangles have 2 long sides and 2 short sides. All the sides in a square are equal.

Level 2.3: False. A square has equal sides *[draws a square]* and a rectangle *[draws rectangle]* doesn't.

The student at Level 2.3 reasons based on properties. He knows that a square has all equal sides, but he still believes, as do most elementary school students, that rectangles cannot have all sides equal. Although his conclusion is incorrect, his reasoning about properties of squares and properties of rectangles is correct, given that he believes that squares are not rectangles. Understanding that squares are rectangles requires Level 3.4 reasoning.

Levels 3.1–3.2: I just don't know. The Rectangle Maker did make squares, but it does not seem like squares are really rectangles. But maybe they are, sort of.

Level 3.3: True maybe, but it does not seem right. A square has all the properties of a rectangle—4 right angles, opposite sides equal.

Levels 3.4–4: True. A square has all the properties of a rectangle—4 right angles, opposite sides equal. So a square must be a rectangle. A square is really a special kind of rectangle.

Glossary

ACUTE ANGLE

An acute angle measures between 0 and 90 *degrees*.

ACUTE TRIANGLE

An acute triangle has all acute angles.

ANGLE

An angle is the union or joining of two rays with a common endpoint.

AXIOM

An axiom is a mathematical statement that is assumed to be true.

AXIOMATIC SYSTEM

In an axiomatic system, statements are considered to be true if they are either axioms or they can be logically deduced or proved from a given set of axioms.

CONGRUENT

Two *shapes* are congruent if they have the exact same shape and size; that is, if one can be placed to fit exactly on top of the other. Two *line segments* are congruent if they have the same length. Two *angles* are congruent if they have the same measure. Two *polygons* are congruent if their corresponding sides and angles are congruent.

DEGREE

A degree is a small unit angle used to measure angles and turns. There are 360 degrees in a full turn or rotation. The symbol for degrees is a small raised circle: 45 degrees = 45°.

DIAGONAL OF A QUADRILATERAL

A diagonal is a line segment joining opposite vertices of a quadrilateral.

EQUILATERAL TRIANGLE

An equilateral triangle has all of its sides the same length.

ISOSCELES TRIANGLE

An isosceles triangle has at least two of its sides the same length.

KITE

A kite is a quadrilateral that is symmetric about at least one of its diagonals.

LINE SEGMENT

A line segment is a continuous straight path between two points. (So a line segment is part of a line.)

LINE

A line is a continuous straight path that extends indefinitely in two opposite directions.

OBTUSE ANGLE

An obtuse angle measures between 90 and 180 degrees.

OBTUSE TRIANGLE

An obtuse triangle has one interior angle that is obtuse.

PARALLELOGRAM

A parallelogram is a quadrilateral that has opposite sides parallel.

PARALLEL

Two lines in a plane are parallel if they do not intersect. Two line segments in a plane are parallel if the lines through them are parallel.

PERPENDICULAR

Two lines in a plane are perpendicular if the angle between them is a right angle. Two line segments are perpendicular if the lines through them are perpendicular.

PATH

A path, or continuous path, is what we get when we continuously move a point through space. For instance, we get a path when we drag a pencil on a piece of paper, without lifting it up at any time during this motion. (A path as defined here is usually called a *curve* in mathematics. However, in formal mathematics, a curve can be straight or can bend, which is confusing to elementary students because they think of curves as bending.)

POINT

A point is a location in space. (*Note:* Lines, line segments, angles, and paths are all infinite sets of points.)

POLYGON

A polygon is a simple, closed continuous path consisting of line segments only. The line segments are called the *sides* of the polygon. The points where the sides intersect are its *vertices* (singular, *vertex*).

QUADRILATERAL

A quadrilateral is a four-sided polygon.

RECTANGLE

A rectangle is a quadrilateral that has opposite sides congruent and parallel, and four right angles.

RHOMBUS

A rhombus is a quadrilateral that has all of its sides the same length.

RIGHT ANGLE

A right angle measures 90 degrees.

RIGHT TRIANGLE

A right triangle has one interior angle that is a right angle.

SCALENE TRIANGLE

All three sides of a scalene triangle are different lengths.

SIDE

A side of a polygon is one of its component line segments.

SIMILAR

Two shapes are similar if they have the exact same shape but not necessarily the same size.

SIMPLE CLOSED PATH

A simple closed path starts and ends at the some point (closed) and does not intersect itself anywhere except its start/end point (simple).

SQUARE

A square is a quadrilateral that has all of its sides congruent and four right angles.

SYMMETRY

A figure is symmetric about a line if its reflection, or flip, about that line is equal to the figure (that is, the reflection lands exactly on top of the original figure).

THEOREM

A geometric theorem is a statement about geometry that has been proved true.

Glossary

TRAPEZOID

In CBA, a trapezoid is a quadrilateral that has at least one set of opposite sides parallel. (In some books, trapezoids have one and only one set of opposite sides parallel.)

TRIANGLE

A triangle is a three-sided polygon.

VERTEX

A vertex of a polygon is the point of intersection of two adjacent sides of a polygon. Vertices (the plural of vertex) occur at the endpoints of the sides of the polygon.

References

Baroody, A. J., & Ginsburg, H. P. 1990. "Children's Learning: A Cognitive View." In R. B. Davis, C. A. Maher, & N. Noddings (Eds.), *Constructivist Views on the Teaching and Learning of Mathematics. Journal for Research in Mathematics.* Education Monograph Number 4. Reston, VA: National Council of Teachers of Mathematics, pp. 51–64.

Battista, M. T. 1998a. *Shape Makers: Developing Geometric Reasoning with the Geometer's Sketchpad.* Berkeley, CA: Key Curriculum Press.

Battista, M. T. 1999. "The Mathematical Miseducation of America's Youth: Ignoring Research and Scientific Study in Education." *Phi Delta Kappan* 80(6): 424–33.

Battista, M. T. 2001a. "How Do Children Learn Mathematics? Research and Reform in Mathematics Education." In Thomas Loveless (Ed.), *The Great Curriculum Debate: How Should We Teach Reading and Math?* Washington, DC: Brookings Press, pp. 42–84.

Battista, M. T. 2001b. "A Research-Based Perspective on Teaching School Geometry." In J. Brophy (Ed.), *Advances in Research on Teaching: Subject-Specific Instructional Methods and Activities.* Advances in Research on Teaching, Vol. 8, JAI Press, pp. 145–85.

Battista, M. T. 2001c. "*Shape Makers*: A Computer Environment that Engenders Students' Construction of Geometric Ideas and Reasoning." *Computers in the Schools* 17(1): 105–20.

Battista, M. T. 2002a. "Learning Geometry in a Dynamic Computer Environment." *Teaching Children Mathematics* 8(1): 333–9.

Battista, M. T. 2002b. "Learning in an Inquiry-Based Classroom: Fifth Graders' Enumeration of Cubes in 3D Arrays." In J. Sowder & B. Schappelle (Eds.), *Lessons Learned from Research.* Reston, VA: National Council of Teachers of Mathematics, pp. 75–84.

Battista, M. T. 2003. *Shape Makers: Developing Geometric Reasoning in the Middle School with the Geometer's Sketchpad.* Berkeley, CA: Key Curriculum Press. [First published in 1998]

Battista, M. T. 2004. "Applying Cognition-Based Assessment to Elementary School Students' Development of Understanding of Area and Volume Measurement." *Mathematical Thinking and Learning* 6(2): 185–204.

Battista, M. T. 2007a. "The Development of Geometric and Spatial Thinking." In F. Lester (Ed.), *Second Handbook of Research on Mathematics Teaching and Learning.* Reston, VA: National Council of Teachers of Mathematics, pp. 843–908.

Battista, M. T. 2007b. "Learning with Understanding: Principles and Processes in the Construction of Geometric Ideas." In M. E. Strutchens & W. G. Martin (Eds.), *69th*

NCTM Yearbook, The Learning of Mathematics. Reston, VA: National Council of Teachers of Mathematics, pp. 65–79.

Battista, M. T. 2008a. "Development of the *Shape Makers* Geometry Microworld: Design Principles and Research." In G. Blume & K. Heid (Eds.), *Research on Technology in the Learning and Teaching of Mathematics: Cases and Perspectives, Vol. 2*. Charlotte, NC: NCTM/Information Age Publishing, pp. 131–56.

Battista, M. T. 2008b. "Representations and Cognitive Objects in Modern School Geometry." In G. Blume & K. Heid (Eds.), *Research on Technology in the Learning and Teaching of Mathematics: Cases and Perspectives, Vol. 2*. Charlotte, NC: NCTM/Information Age Publishing, pp. 341–62.

Battista, M. T. 2009. "Highlights of Research on Learning School Geometry." In T. Craine & R. Rubenstein (Eds.), *2009 Yearbook, Understanding Geometry for a Changing World*. Reston, VA: National Council of Teachers of Mathematics.

Battista, M. T., & Clements, D. H. 1995. "Geometry and Proof." (Connecting Research to Teaching). *Mathematics Teacher* 88(1): 48–54.

Battista, M. T., & Clements, D. H. 1996. "Students' Understanding of Three-Dimensional Rectangular Arrays of Cubes." *Journal for Research in Mathematics Education* 27(3): 258–92.

Battista, M. T., Clements, D. H., Arnoff, J., Battista, K., & Borrow, C. V. A. 1998. "Students' Spatial Structuring and Enumeration of 2D Arrays of Squares." *Journal for Research in Mathematics Education* 29(5): 503–32.

Black, P., & Wiliam, D. 1998. "Raising Standards Through Classroom Assessment." *Phi Delta Kappan* 80(2): 139–48.

Borrow, Caroline. 2000. *An Investigation of the Development of 6th-Grade Students' Geometric Reasoning and Conceptualizations of Geometric Polygons in a Computer Microworld.* Unpublished doctoral dissertation, Kent State University.

Bransford, J. D., Brown, A. L., & Cocking, R. R. 1999. *How People Learn: Brain, Mind, Experience, and School*. Washington, DC: National Research Council.

Burger, William F., & Shaughnessy, J. Michael. 1986. "Characterizing the van Hiele Levels of Development in Geometry." *Journal for Research in Mathematics Education* 17(January): 31–48.

Buschman, Larry. 2001. "Using Student Interviews to Guide Classroom Instruction: An Action Research Project." *Teaching Children Mathematics* (December): 222–27.

Carpenter, T. P., & Fennema, E. 1991. "Research and Cognitively Guided Instruction." In E. Fennema, T. P. Carpenter, & S. J. Lamon (Eds.), *Integrating Research on Teaching and Learning Mathematics*. Albany: State University of New York Press, pp. 1–16.

Carpenter, T. P., Franke, M. L., Jacobs, V. R., Fennema, E., & Empson, S. B. 1998. "A Longitudinal Study of Invention and Understanding in Children's Multidigit Addition and Subtraction." *Journal for Research in Mathematics Education* 29(1): 3–20.

Clements, D. H., & Battista, M. T. 1992. "Geometry and Spatial Reasoning." In D. Grouws (Ed.), *Handbook of Research on Mathematics Teaching and Learning*. New York: NCTM/Macmillan, pp. 420–64.

Cobb, P., & Wheatley, G. 1988. "Children's Initial Understanding of Ten." *Focus on Learning Problems in Mathematics* 10(3): 1–28.

Cobb, P., Wood, T., Yackel, E., Nicholls, J., Wheatley, G., Trigatti, B., & Perlwitz, M. 1991. "Assessment of a Problem-Centered Second-Grade Mathematics Project." *Journal for Research in Mathematics Education* 22(1): 3–29.

De Corte, E., Greer, B., & Verschaffel, L. 1996. "Mathematics Teaching and Learning." In D. C. Berliner & R. C. Calfee (Eds.), *Handbook of Educational Psychology*. New York: Simon & Schuster Macmillan, pp. 491–549.

de Villiers, Michael. 2003. *Rethinking Proof with The Geometer's Sketchpad*. Emeryville, CA: Key Curriculum Press.

Fennema, E., Carpenter, T. P., Franke, M. L., Levi, L., Jacobs, V. R., & Empson, S. B. 1996. "A Longitudinal Study of Learning to Use Children's Thinking in Mathematics Instruction." *Journal for Research in Mathematics Education* 27(4): 403–34.

Fennema, E., & Franke, M. L. 1992. "Teachers' Knowledge and Its Impact." In D. A. Grouws (Ed.), *Handbook of Research on Mathematics Teaching*. Reston, VA: National Council of Teachers of Mathematics/Macmillan, pp. 127–64.

Fuson, K. C., Wearne, D., Hiebert, J. C., Murray, H. G., Human, P. G., Olivier, A. L., et al. 1997. "Children's Conceptual Structures for Multidigit Numbers and Methods of Multidigit Addition and Subtraction." *Journal for Research in Mathematics Education* 28(2): 130–62.

Fuys, D., Geddes, D., & Tischler, R. 1988. *The van Hiele Model of Thinking in Geometry Among Adolescents. Journal for Research in Mathematics Education*. Monograph No. 3. Reston, VA: National Council of Teachers of Mathematics.

Greeno, J. G. 1991. "Number Sense as Situated Knowing in a Conceptual Domain." *Journal for Research in Mathematics Education* 22(3): 170–218.

Greeno, J. G., Collins, A. M., & Resnick, L. 1996. "Cognition and Learning." In D. C. Berliner & R. C. Calfee (Eds.), *Handbook of Educational Psychology*. New York: Simon & Schuster Macmillan, pp. 15–46.

Hiebert, J., & Carpenter, T. P. 1992. "Learning and Teaching with Understanding." In D. A. Grouws (Ed.), *Handbook of Research on Mathematics Teaching*. Reston, VA: National Council of Teachers of Mathematics/Macmillan, pp. 65–97.

Hoyles, C., & Jones, K. 1998. "Proof in Dynamic Geometry Contexts." In C. Mammana & V. Villani (Eds.), *Perspectives on the Teaching of Geometry for the 21st Century*. Dordrecht, The Netherlands: Kluwer, pp. 121–28.

Johnson-Laird, P. N. 1983. *Mental Models: Towards a Cognitive Science of Language, Inference, and Consciousness*. Cambridge, MA: Harvard University Press.

Johnson-Laird, P. N. 1998. "Imagery, Visualization, and Thinking." In J. Hochberg (Ed.), *Perception and Cognition at Century's End*. San Diego, CA: Academic Press, pp. 441–67.

Jones, Keith. 2000. "Providing a Foundation for Deductive Reasoning: Students' Interpretations When Using Dynamic Geometry Software and Their Evolving Mathematical Explanations." *Educational Studies in Mathematics* 44(1–2): 55–85.

Laborde, C. 1998. "Visual Phenomena in the Teaching/Learning of Geometry in a Computer-Based Environment." In C. Mammana & V. Villani (Eds.), *Perspectives on the Teaching of Geometry for the 21st Century*. (Dordrecht, The Netherlands: Kluwer, pp. 113–21.

Lester, F. K. 1994. "Musing About Mathematical Problem-Solving Research: 1970–1994." *Journal for Research in Mathematics Education* 25(6): 660–75.

National Council of Teachers of Mathematics (NCTM). 2000. *Principles and Standards for School Mathematics*. Reston, VA: NCTM.

National Research Council. 1989. *Everybody Counts*. Washington, DC: National Academy Press.

Olive, J. 1998. "Opportunities to Explore and Integrate Mathematics with *The Geometer's Sketchpad*." In R. Lehrer & D. Chazan (Eds.), *Designing Learning Environments for Developing Understanding of Geometry and Space*. Mahwah, NJ: Lawrence Erlbaum, pp. 395–418.

Pinker, S. 1997. *How the Mind Works*. New York: W. W. Norton.

Prawat, R. S. 1999. "Dewey, Peirce, and the Learning Paradox." *American Educational Research Journal* 36(1): 47–76.

Romberg, T. A. 1992. "Further Thoughts on the Standards: A Reaction to Apple." *Journal for Research in Mathematics Education* 23(5): 432–7.

Schoenfeld, A. C. 1994. "What Do We Know About Mathematics Curricula." *Journal of Mathematical Behavior* 13: 55–80.

Serra, M. 2003. *Discovering Geometry: An Investigative Approach*. Emeryville, CA: Key Curriculum Press.

Steffe, L. P. 1988. "Children's Construction of Number Sequences and Multiplying Schemes." In J. Hiebert & M. Behr (Eds.), *Number Concepts and Operations in the Middle Grades*. Reston, VA: National Council of Teachers of Mathematics, pp. 119–40.

Steffe, L. P. 1992. "Schemes of Action and Operation Involving Composite Units." *Learning and Individual Differences* 4(3): 259–309.

Steffe, L. P., & D'Ambrosio, B. S. 1995. "Toward a Working Model of Constructivist Teaching: A Reaction to Simon." *Journal for Research in Mathematics Education* 26(2): 146–59.

Steffe, L. P., & Kieren, T. 1994. "Radical Constructivism and Mathematics Education." *Journal for Research in Mathematics Education* 25(6): 711–33.

Usiskin, Z., Hirschorn, D., & Coxford, A. 2003. *Geometry* (University of Chicago School Mathematics Project). New York: McGraw-Hill.

van Hiele, P. M. 1986. *Structure and Insight*. Orlando, FL: Academic Press.

von Glasersfeld, Ernst. 1995. *Radical Constructivism: A Way of Knowing and Learning*. Washington, DC: Falmer Press.

Yerushalmy, M., & Chazan, D. (1993). "Overcoming Visual Obstacles with the Aid of the Supposer." In J. L. Schwartz, M. Yerushalmy, & B. Wilson (Eds.), *The Geometric Supposer: What Is It a Case Of?* Hillsdale, NJ: Lawrence Erlbaum, pp. 25–56.